Springer Series in

OPTICAL SCIENCES 86

founded by H.K.V. Lotsch

Springer
Berlin
Heidelberg
New York
Hong Kong
London
Milan
Paris
Tokyo

Springer Series in

OPTICAL SCIENCES

The Springer Series in Optical Sciences, under the leadership of Editor-in-Chief *William T. Rhodes*, Georgia Institute of Technology, USA, and Georgia Tech Lorraine, France, provides an expanding selection of research monographs in all major areas of optics: lasers and quantum optics, ultrafast phenomena, optical spectroscopy techniques, optoelectronics, quantum information, information optics, applied laser technology, industrial applications, and other topics of contemporary interest.
With this broad coverage of topics, the series is of use to all research scientists and engineers who need up-to-date reference books.

The editors encourage prospective authors to correspond with them in advance of submitting a manuscript. Submission of manuscripts should be made to the Editor-in-Chief or one of the Editors. See also http://www.springer.de/phys/books/optical_science/

Motoichi Ohtsu (Ed.)

Progress in Nano-Electro-Optics I

Basics and Theory of Near-Field Optics

With 118 Figures

Springer

Professor Dr. Motoichi Ohtsu
Tokyo Institute of Technology
Interdisciplinary Graduate School
of Science and Engineering
4259 Nagatsuta-cho, Midori-ku
Yokohama 226-8502
Japan
E-mail: ohtsu@ae.titech.ac.jp

ISSN 0342-4111

ISBN 3-540-43504-2 Springer-Verlag Berlin Heidelberg New York

Library of Congress Cataloging-in-Publication Data

Progress in nano-electro-optics I : basics and theory of near-field optics / Motoichi Ohtsu (ed.).
p.cm. – (Springer series in optical siences ; v. 86)
Includes bibliographical references and index.
ISBN 3540435042 (alk. paper)
1. Electrooptics. 2. Nanotechnology. 3. Near-field microscopy. I. Ohtsu, Motoichi. II. Series.
TA1750 .P75 2002   621.381'045–dc21   2002030321

Springer-Verlag Berlin Heidelberg New York
a member of BertelsmannSpringer Science+Business Media GmbH

http://www.springer.de

Printed in Germany

Camera-ready by the author using a Springer TEX macropackage
Cover concept by eStudio Calamar Steinen using a background picture from The Optics Project. Courtesy of John T. Foley, Professor, Department of Physics and Astronomy, Mississippi State University, USA.
Cover production: *design & production* GmbH, Heidelberg

Printed on acid-free paper   SPIN 10864333   57/3141/di   5 4 3 2 1 0

Preface to *Progress in Nano-Electro-Optics*

Recent advances in electro-optical systems demand drastic increases in the degree of integration of photonic and electronic devices for large-capacity and ultrahigh-speed signal transmission and information processing. Device size has to be scaled down to nanometric dimensions to meet this requirement, which will become even more strict in the future. In the case of photonic devices, this requirement cannot be met only by decreasing the sizes of materials. It is indispensable to decrease the size of the electromagnetic field used as a carrier for signal transmission. Such a decrease in the size of the electromagnetic field beyond the diffraction limit of the propagating field can be realized in optical near fields.

Near-field optics has progressed rapidly in elucidating the science and technology of such fields. Exploiting an essential feature of optical near fields, i.e., the resonant interaction between electromagnetic fields and matter in nanometric regions, important applications and new directions such as studies in spatially resolved spectroscopy, nano-fabrication, nano-photonic devices, ultrahigh-density optical memory, and atom manipulation have been realized and significant progress has been reported. Since nano-technology for fabricating nanometric materials has progressed simultaneously, combining the products of these studies can open new fields to meet the above-described requirements of future technologies.

This unique monograph series entitled "Progress in Nano-Electro-Optics" is being introduced to review the results of advanced studies in the field of electro-optics at nanometric scales and covers the most recent topics of theoretical and experimental interest on relevant fields of study (e.g., classical and quantum optics, organic and inorganic material science and technology, surface science, spectroscopy, atom manipulation, photonics, and electronics). Each chapter is written by leading scientists in the relevant field. Thus, high-quality scientific and technical information is provided to scientists, engineers, and students who are and will be engaged in nano-electro-optics and nano-photonics research.

I gratefully thank the members of the editorial advisory board for valuable suggestions and comments on organizing this monograph series. I wish to express my special thanks to Dr. T. Asakura, Editor of the Springer Series in Optical Sciences, Professor Emeritus, Hokkaido University for recommending me to publish this monograph series. Finally, I extend an acknowledgement to Dr. Claus Ascheron of Springer-Verlag, for his guidance and suggestions, and to Dr. H. Ito, an associate editor, for his assistance throughout the preparation of this monograph series.

Yokohama, October 2002 *Motoichi Ohtsu*

Preface to Volume I

This volume contains five review articles focusing on various aspects of nano-electro-optics. The first article deals with fiber probes and related devices for generating and detecting optical near fields with high efficiency and resolution. These are essential tools for studying and applying optical near fields, and thus the article is most appropriate as the first chapter in this monograph series.

The second article is devoted to modulation of an electron beam by optical near fields. It is an important study on fundamental physical processes involving electron–light interaction, including quantum effects. This interaction is related to the well known Smith–Purcell and Schwarz–Hora effects.

The third article concerns fluorescence spectroscopy. To excite sample molecules, an evanescent surface plasmon field close to metallic surfaces is utilized, which is becoming an important and popular technique in studying nanometric species such as dye molecules.

The article that follows describes the spatially resolved near-field photoluminescence spectroscopy of semiconductor quantum dots, which will become a essential component in future electro-optical devices and systems. This spectroscopy also provides high spectral resolution, and thus it is a powerful tool for collecting precise information about a single quantum dot with a photoluminescence linewidth an order of magnitude narrower than that of ensemble quantum dots.

The last article deals with the quantum theory of optical near fields based on the concepts of a projection operator and an exciton–polariton. Since this theory concerns the optical near-field system coupling nanometric and macroscopic systems, it accounts for all the essential features of interaction between optical near fields and nano-matter, atoms, and molecules. Further, it can also be utilized to design future nano-electro-optical devices and systems.

I hope that this volume will be a valuable resource for readers and future specialists.

Yokohama, October 2002 *Motoichi Ohtsu*

Preface to Volume I

Contents

List of Contributors

Yasuhiko Arakawa
Institute of Industrial Science
University of Tokyo
4-6-1 Komaba, Meguro-ku
Tokyo 153-8505, Japan
arakawa@iis.u-tokyo.ac.jp

Jongsuck Bae
Research Institute
of Electrical Communication
Tohoku University
2-1-1 Katahira, Aoba-ku
Sendai 980-8577, Japan
bae@riec.tohoku.ac.jp

Ryo Ishikawa
Research Institute
of Electrical Communication
Tohoku University
2-1-1 Katahira, Aoba-ku
Sendai 980-8577, Japan
issi@riec.tohoku.ac.jp

Wolfgang Knoll
Max-Planck-Institut
für Polymerforschung
Ackermannweg 10
55128 Mainz, Germany
knoll@mpip-mainz.mpg.de

Kiyoshi Kobayashi
ERATO Localized Photon Project
Japan Science and Technology
Corporation
687-1 Tsuruma
Machida, Tokyo 194-0004, Japan
kkoba@ohtsu.jst.go.jp

Max Kreiter
Max-Planck-Institut
für Polymerforschung
Ackermannweg 10
55128 Mainz, Germany
kreiter@mpip-mainz.mpg.de

Koji Mizuno
Research Institute
of Electrical Communication
Tohoku University
2-1-1 Katahira, Aoba-ku
Sendai 980-8577, Japan
koji@riec.tohoku.ac.jp

Thomas Neumann
Max-Planck-Institut
für Polymerforschung
Ackermannweg 10
55128 Mainz, Germany
neumann@mpip-mainz.mpg.de

Motoichi Ohtsu
Interdisciplinary Graduate School
of Science and Technology
Tokyo Institute of Technology
4259 Nagatsuta-cho, Midori-ku
Yokohama 226-8502, Japan
ohtsu@ae.titech.ac.jp

Suguru Sangu
ERATO Localized Photon Project
Japan Science and Technology
Corporation
687-1 Tsuruma
Machida, Tokyo 194-0004, Japan
sangu@ohtsu.jst.go.jp

Yasunori Toda
Graduate School of Engineering
Hokkaido University
North 13-West 8, Sapporo
Hokkaido 060-8628, Japan
toda@eng.hokudai.ac.jp

Takashi Yatsui
ERATO Localized Photon Project
Japan Science and Technology
Corporation
687-1 Tsuruma
Machida, Tokyo 194-0004, Japan
yatsui@ohtsu.jst.go.jp

High-Throughput Probes for Near-Field Optics and Their Applications

T. Yatsui and M. Ohtsu

1 High-Throughput Probes

Recent developments in near-field optical microscopy have made it possible to obtain optical images with nanometer-scale spatial resolution by scanning a fiber probe with a sub-wavelength aperture [1]. In attempts to improve performance in spatially resolved spectroscopy, a serious problem of the fiber probe is its low throughput (in the case of illumination-mode operation, the throughput is defined as the ratio of the output light power at the apex to the incident light power coupled into the fiber). The essential cause of the low throughput is the guiding loss along the metallized tapered core. Based on a mode analysis of the tapered core, we review our work to realize high-throughput probes.

1.1 Mode Analysis in a Metallized Tapered Probe

Mode analysis has been carried out by approximating a tapered core as a concatenated cylindrical core with a metal cladding (see Fig. 1) [2].

Figure 2 shows the equivalent refractive indices of relevant modes as a function of a core diameter D (at $\lambda = 830$ nm), which is derived by a mode analysis for an infinitely thick gold coated core. Refractive indices of the glass and the gold used for this derivation are 1.53 and $0.17 + i5.2$ [3], respectively. Definitions of the EH_{11} and HE_{11} modes in this figure are based on those in [2]. This figure shows that the cut-off core diameter (D_c) of HE_{11} mode is

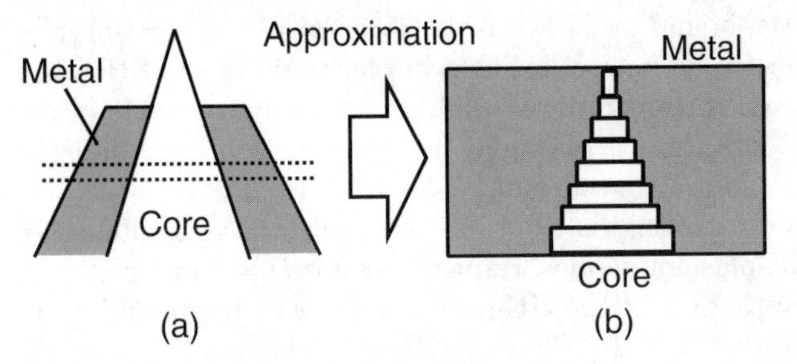

Fig. 1. (**a**) Metal coated tapered core. (**b**) Concatenated cylindrical core

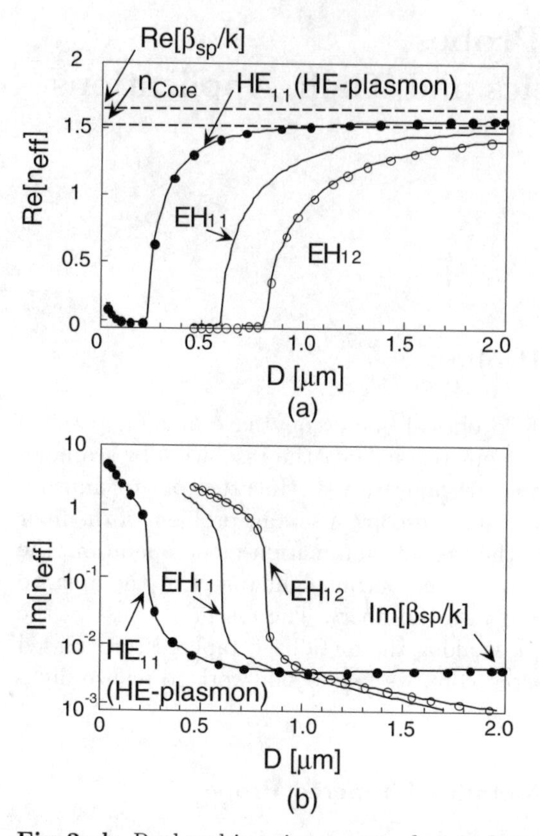

Fig. 2a,b. Real and imaginary part of equivalent refractive indices of the relevant modes guided through a glass core with a gold metallic film, respectively. D, core diameter; β_{sp}, propagation constant of the surface plasmon at a planar boundary between gold and air

as small as 30 nm, while that of the EH_{11} is 450 nm. It means that only the HE_{11} mode can excite the optical near field efficiently in $D < 100$ nm. This figure also shows that, as D increases, the equivalent refractive indices of the EH_{11} and HE_{11} modes approach the refractive index of the core and that of a surface plasmon at the boundary between the plane surface of semi-infinite gold and the air, respectively. It also means that the origin of the EH_{11} and HE_{11} modes in a tapered core with metal cladding is the lowest mode in the single-mode fiber and the surface plasmon, respectively. Thus, we call the HE_{11} the HE-plasmon modes from now on.

Figure 3a–c shows the spatial distributions of the electric field intensities of EH_{11}, EH_{12}, and HE-plasmon modes, respectively, for $D = 3$ μm. Note the localization of the electric field of the HE-plasmon mode at the boundary of the core and metal cladding (Fig. 3c). Thus, the HE-plasmon mode should not be easily excited in the conventional core because of its low coupling efficiency

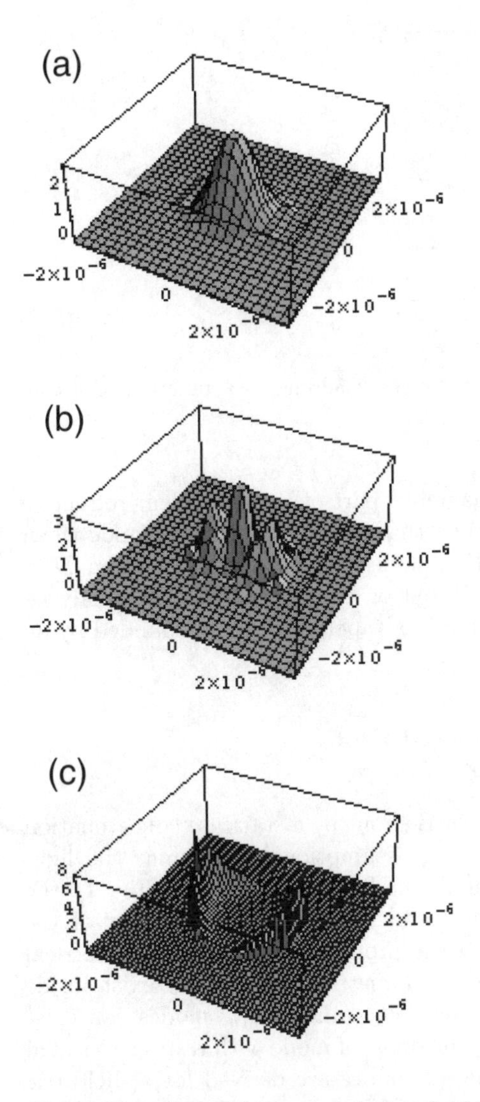

Fig. 3. Spatial distributions of electric field intensity for $D = 3\,\mu\mathrm{m}$. (**a**) EH_{11} mode, (**b**) EH_{12} mode, and (**c**) HE-plasmon mode

with the lowest-order mode (HE_{11}) guided through the optical fiber, owing to the mode mismatch between the HE_{11} and HE-plasmon modes at the foot of the tapered core. As a consequence, the throughput of the conventional fiber probe is very low.

The skin depth in the metal cladding has been estimated by considering the EH_{11} mode as a guided mode. Figure 4 shows the skin depth in the metal cladding as a function of D. Note that the light hardly leaks into the

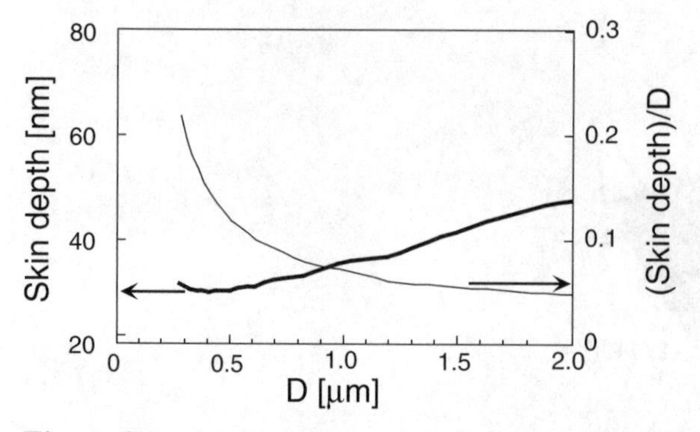

Fig. 4. Skin depth of EH_{11} mode in the metal cladding as a function of a core diameter (D)

cladding. Furthermore, since the imaginary part of the equivalent refractive index of the EH_{11} mode is smaller than that of the HE-plasmon mode when $D > 800$ nm (see Fig. 2b), the EH_{11} mode should be excited when $D >$ 800 nm for increasing throughput. Based on this calculation, we analyzed the characteristics of light propagation in a tapered probe surrounded by an ideal metal in the next section.

1.2 Light Propagation in a Tapered Probe with Ideal Metal Cladding

Light propagation analysis has been carried out by a staircase concatenation method. This is based on the following assumptions; first, since the light hardly penetrates the metal cladding in the large D region (see Fig. 4), the cladding is assumed to be an ideal metal, i.e., the imaginary part of the refractive index is zero. Second, since the propagation mode is symmetrical in the single-mode optical fiber, only symmetrical modes can exist inside this probe. By these assumptions, only TE_{1m} and TM_{1m} modes can exist inside the probe, where m represents the order of mode with respect to radial direction. For this analysis, the relevant modes are derived for a dielectric core ($n = 1.53$) coated with an infinitely thick ideal metal. Figure 5a–c shows the spatial distribution of electric field intensities of TE_{11}, TE_{12}, and TM_{11} modes, respectively. Figure 5d shows the equivalent refractive indices of these modes.

Using these modes, the spatial distribution of the electric field intensity was calculated by a staircase concatenation method for a tapered probe with a cone angle (α) of 25°. For this calculation, a Gaussian incident light beam with full width at half maximum (FWHM) of 3 μm is assumed, where the foot core diameter is 3 μm. Figure 6 represents the calculated result, which shows that the spatial distribution of the electric field intensity goes through the progression "single peak → double peak → triple peak → single peak".

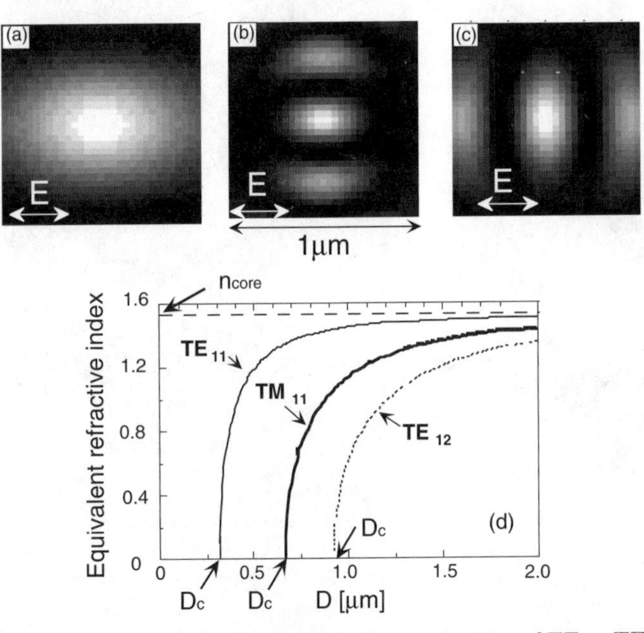

Fig. 5. (**a–c**) Spatial distribution of the intensities of TE_{11}, TE_{12}, and TM_{11} modes, respectively, in a dielectric core ($n = 1.53$) with an ideal metal cladding ($\lambda = 830$ nm). The vector E in these figures represents the direction of incident light polarization. (**d**) Equivalent refractive indices of the three modes. D_c is the cut-off diameter

Figure 7 shows the intensity at the center of the core and the FWHM of the central part as a function of D. It shows that the locally maximum electric field intensities are realized at $D = 1.47$ μm, 920 nm, and 315 nm and that the smallest spots are realized at $D = 1.47$ μm and 920 nm. Note that the minimum of FWHM is as small as 175 nm ($\sim \lambda/5$) at $D = 920$ nm.

Figure 8 shows the calculated electric-field intensity for relevant modes at the center of the core. Note that all electric-field intensities peak at each D_c due to the resonance. Furthermore, as has been shown by Fig. 7, the FWHM takes its minimum at each D_c except at $D = D_c$ of the TE_{11} mode, where only the TE_{11} mode can exist. The decrease in the FWHM is due to the interference of the guided modes inside the metallized tapered probe. Thus, we call these dependencies of electric-field intensity and FWHM on the core diameter D as the "interference characteristics of the guided modes" from now on.

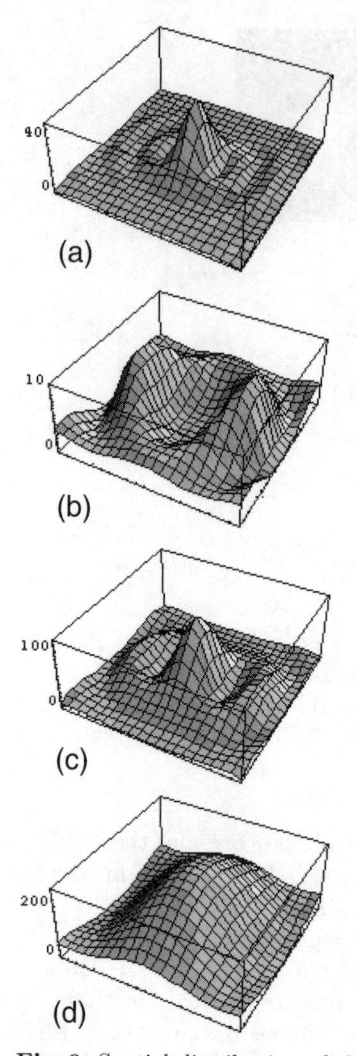

Fig. 6. Spatial distribution of electric-field intensity in a tapered probe with $\alpha = 25°$ calculated by a staircase concatenation method. (**a**) $D = 1.47$ µm, (**b**) $D = 1.06$ µm, (**c**) $D = 920$ nm, and (**d**) $D = 315$ nm

1.3 Measurement of the Spatial Distribution of Optical Near-Field Intensity in the Tapered Probe

High throughput and a small spot can be expected by optimizing the above-mentioned interference characteristics of the guided modes. In this section, we describe experimental results which show the validity to this expectation.

For this purpose, the optical near field generated on the apertured fiber probe was observed by scanning another probe with $D = 100$ nm over the aperture to scatter and detect the optical near field [4]. The experimental

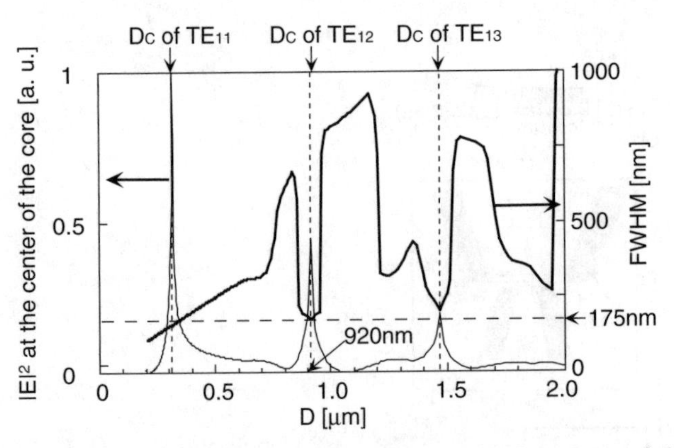

Fig. 7. Calculated electric field intensity at the center of the core and the FWHM of the profiles given in Fig. 6. D_c is the cut-off diameter

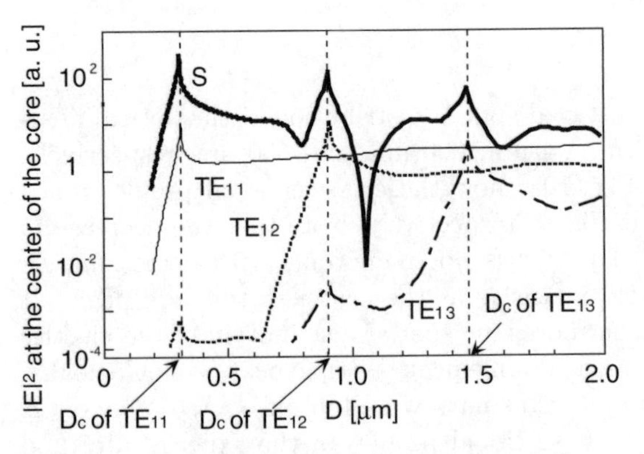

Fig. 8. Calculated electric-field intensity at the center of the core. S, $|\sum E_i|^2$. D_c is the cut-off diameter

setup, i.e., probe-to-probe method, is shown in Fig. 9, where the separation between the probes was kept within several nanometers by shear-force techniques.

Linearly polarized light from a laser diode (830 nm) was coupled into the single-tapered apertured probe, which was fabricated in the following three steps:

1. The GeO_2-doped core with a diameter of 3 μm was tapered with $\alpha = 25°$ by the selective chemical etching technique [5].
2. The core was coated with 500-nm-thick gold film.
3. Since the core diameter D cannot be defined accurately by the angled evaporation [6], the apex of the core was removed by a focused ion beam (FIB) to form a definite aperture. The value of D was then determined from scanning electron micrographic images.

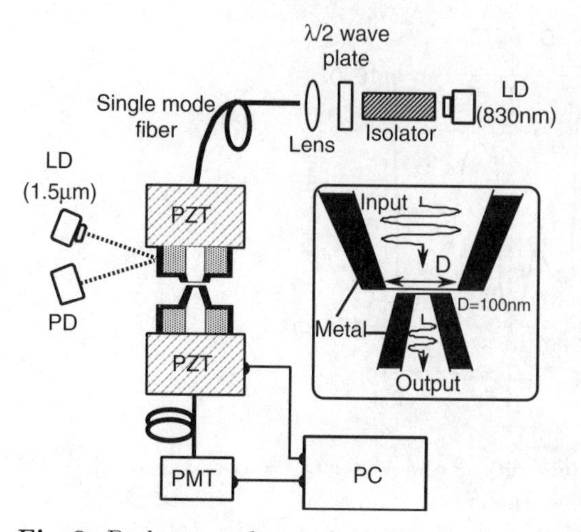

Fig. 9. Probe-to-probe method. LD, laser diode; PD, photo diode; PMT, photomultiplier tube

Figure 10a–d shows the observed spatial distribution of the optical near-field intensity with D of 2 μm, 1.4 μm, 900 nm, and 550 nm, respectively. Curves A, B, C, and D in Fig. 10e show the cross-sectional profiles along the dashed white line in Fig. 10a–d, respectively. Note that the decrease in FWHM of the cross-sectional profile is not monotonic with decreasing D, and that the optical near-field intensity at the center of the core peaks at $D = 900$ nm (curve C). Furthermore, the spatial distribution of the electric field goes through the progression "double peak → triple peak → single peak". Note that the FWHM of curve C is as narrow as 150 nm($< \lambda/5$). Curves A and B in Fig. 11 show the cross-sectional profiles of the experimental and calculated results, respectively. Note that curve B is in good agreement with curve A with respect to the FWHM of the main lobe, the position of the side peak, and the ratio of side peak intensity to the central peak intensity. In Fig. 12, the measured intensity at the center of the core is plotted as a function of D. This is in good agreement with calculated results with respect to the maximum intensity around $D = D_c$ (of TE_{12} mode).

In Fig. 13, the throughput is plotted as a function of D, where the throughput is defined as (output power)/(input power coupled into the fiber). This figure shows that the probe with $D = 900$ nm has throughput of 10 %, 1000 times that of the probe with $D = 150$ nm. Using a fiber probe with $D = D_c$ (of TE_{12} mode), we obtained both high throughput (10 %) and a small spot (150 nm). Because of the guiding loss along the metallized tapered core, the throughput decreases drastically when $D < D_c$ (of the TE_{11} mode).

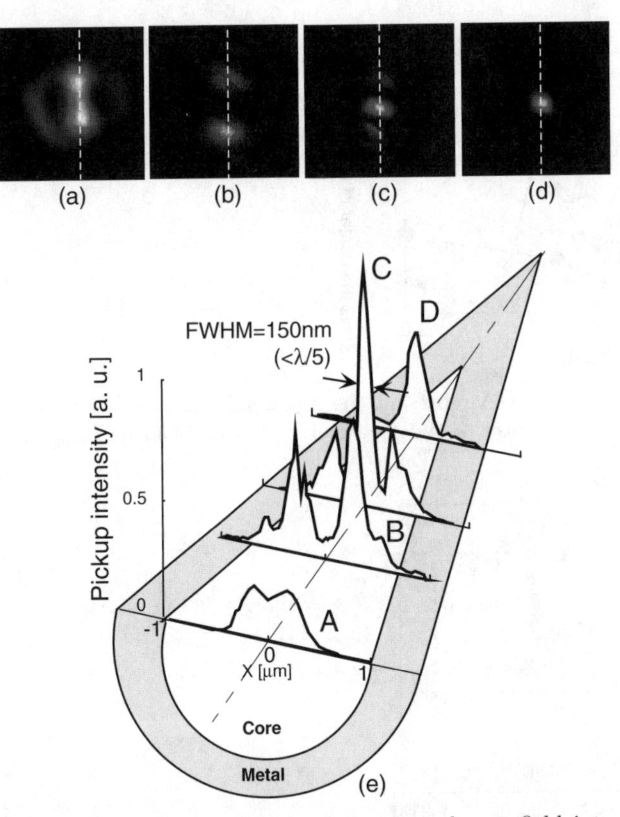

Fig. 10. Spatial distribution of the optical near-field intensity in a conventional single-tapered probe. (**a**–**d**) are for a single-tapered probe with $D = 2$ μm, 1.4 μm, 900 nm, and 550 nm, respectively. Image sizes are 1.5×1.5 μm. Curves A, B, C, and D in (**e**) are cross-sectional profiles along the dashed white line in (**a**), (**b**), (**c**), and (**d**) respectively

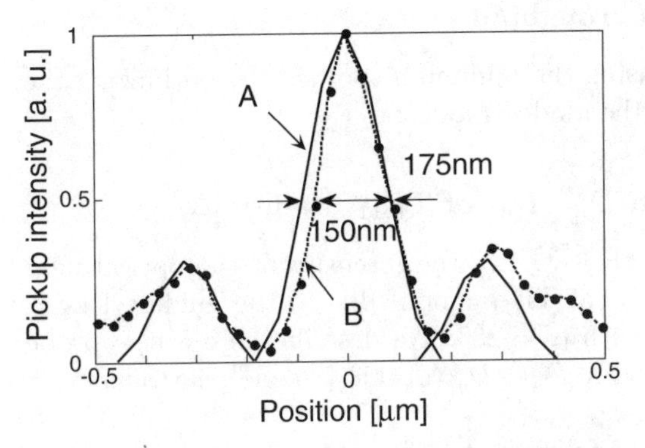

Fig. 11. Spatial distribution at $D = 900$ nm. Curve A: calculated result; curve B: experimental result

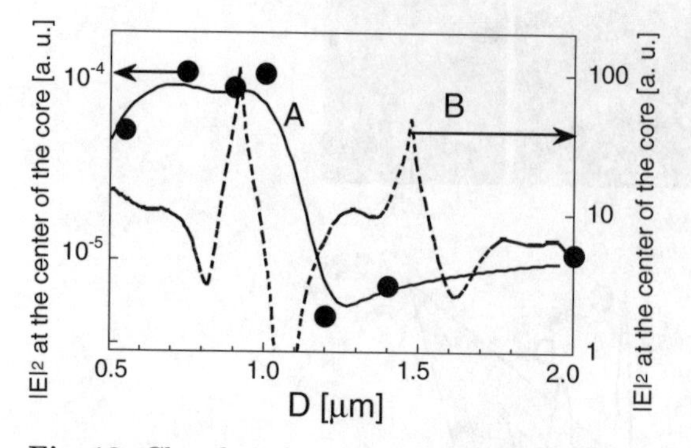

Fig. 12. Closed circles and curve A represent the measured relations between the electric-field intensity at the center of the core and D. Curve B is the calculated result

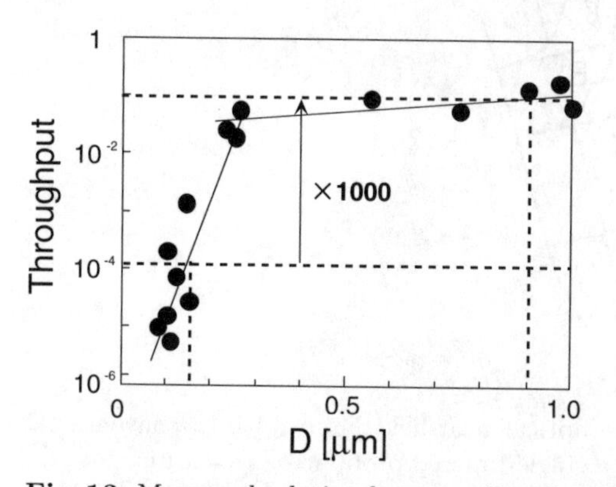

Fig. 13. Measured relation between D and the throughput for conventional single-tapered apertured probe

1.4 Further Increase in Throughput

We review our work on decreasing the minimum spot size by optimizing the interference characteristics of the guided modes.

Triple-Tapered Probe with $D = D_c$ (of TE_{11} Mode)

As described in the last part of Sect. 1.3, the intensity of the optical near field does not peak at $D = D_c$ (of TE_{11} mode) due to the guiding loss in the metallized tapered probe with $\alpha = 25°$. We describe here a new probe with short probe length and with $D = D_c$ (of TE_{11} mode), introducing a triple-tapered structure [7].

The triple-tapered probe was fabricated in five steps:

1. The GeO_2-doped core was tapered with $\alpha = 25°$ by the selective chemical etching technique [5].

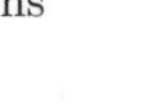

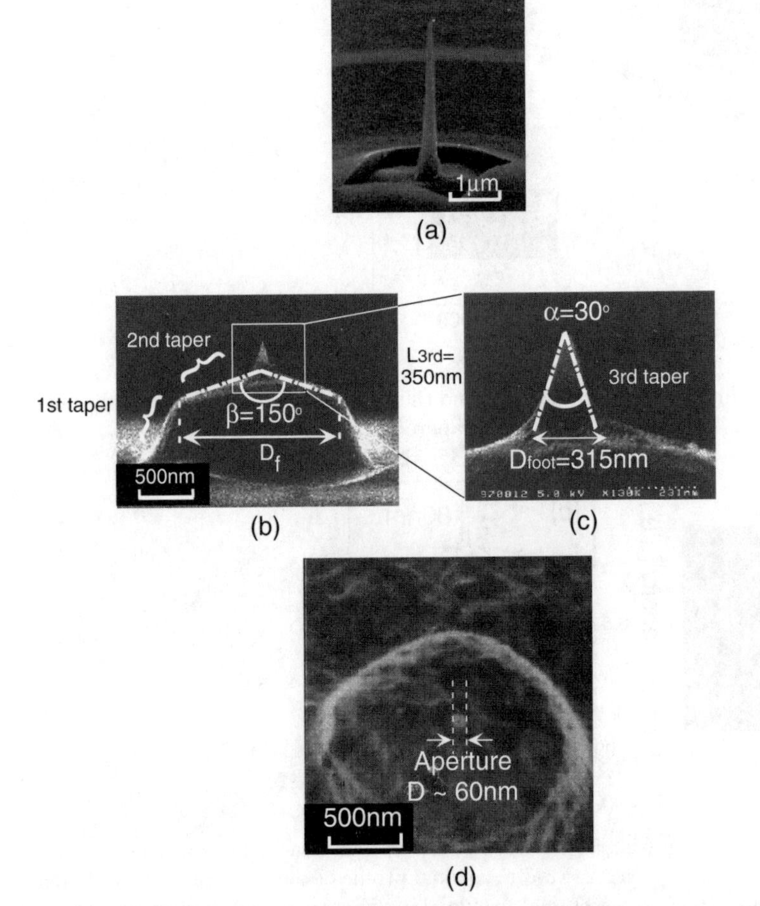

Fig. 14. SEM images of triple-tapered probe. (**a**) Results of step 2. (**b**), (**c**) Profile of a triple-tapered core formed by step 3 and a magnified image of the third taper, respectively. (**d**) Bird's-eye view of the probe after steps 5

2. To reduce the core diameter, the FIB was irradiated over the sharpened core (Fig. 14a).
3. The first and second tapers of the core were formed by the chemical etching technique. The foot diameter of second taper D_f was 1.6 μm (Fig. 14b). The length and apex diameter of the third tapers were 315 nm, and 25 nm, respectively (Fig. 14c). The cone angles of the second and third tapers were 30° and 150°, respectively.
4. The core was coated with 300-nm-thick gold film.
5. The top of the third taper was removed by FIB to form an aperture.

Figure 14d shows an SEM image of a fabricated triple-tapered probe with $D = 60$ nm. Since the length of the third taper (350 nm) is much less than

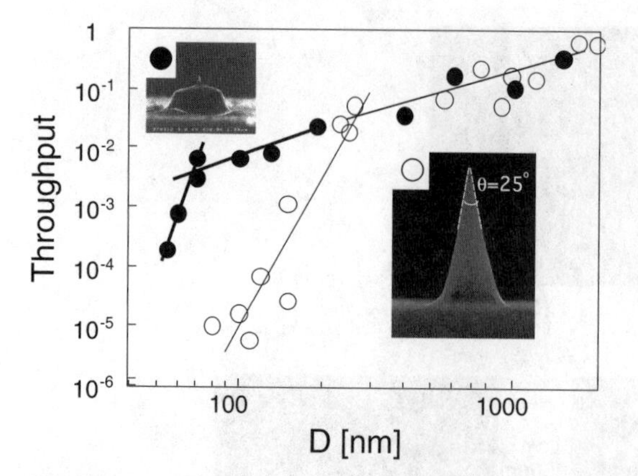

Fig. 15. Measured relations between D and the throughput of a conventional single-tapered probe (*open circles*) and triple-tapered probe (*closed circles*)

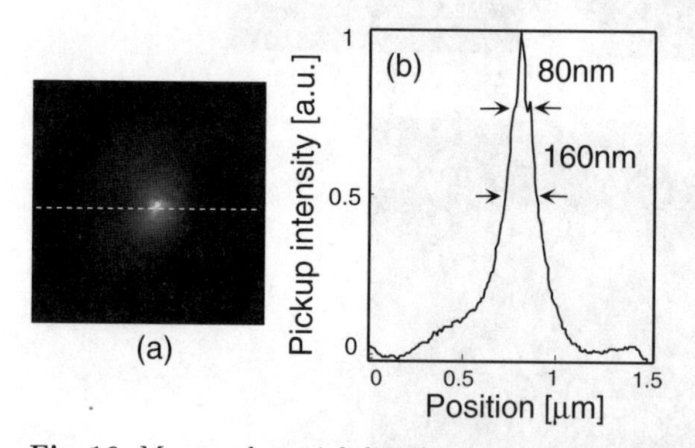

Fig. 16. Measured spatial distribution of the optical near field of a triple-tapered probe with $D = 60$ nm. (**a**) A two-dimensional profile of the distribution. The image is 1.5×1.5 μm. (**b**) Cross-sectional profile along the dashed white line in (**a**)

that of the single-tapered probe (5 μm), a decrease in the propagation loss is expected.

In Fig. 15, the throughput is plotted as a function of D for a triple-tapered probe and a conventional single-tapered probe. It shows that triple-tapered apertured probes have 1000 times the throughput of a single-tapered probe with $D < 100$ nm. Such a drastic increase in throughput of a triple-tapered probe is a consequence of the efficient excitation of an HE-plasmon mode. The efficient excitation of the HE-plasmon mode is attributed to the drastic change in the shape of the probe at $D = D_c$ (of the TE_{11} mode).

We checked whether the triple-tapered structure led to high spatial resolution. For this purpose, the spatial distribution of the optical near-field intensity in a triple-tapered probe with $D = 60$ nm was observed using the probe-to-probe method (see Fig. 9). To enhance the efficiency of light scattering, a sharpened probe with 50-nm-thick gold was used as a scanning probe [8]. Figure 16a and b shows the observed spatial distribution of the optical near-field intensity on the aperture, and its cross-sectional profile, respectively. Due to the thin coating of the metallized probe used for the measurement, the 160-nm FWHM is a consequence of leakage of the propagating far field. However, a very sharp and narrow (80 nm) part at the center of the curve can be attributed to localization of the optical near field due to small α of the third-taper. This part leads to high spatial resolution in imaging, spectroscopy, fabrication, and so on.

Edged Probe for Efficient Excitation of HE-plasmon Mode

An effecient way of exciting the HE-plasmon mode is to utilize coupling of the plasmon by scattering at the edge of the metal [9]. If the tapered core has a sharp edge at its foot, a part of the guided light inside the single-mode fiber can be scattered at this edge and converted to the HE-plasmon mode [10]. We call a probe with such a core an edged probe [4].

Figure 17 shows scanning electron micrographs of a fabricated edged probe. The arrow R in Fig. 17a indicates the direction parallel to the surface from which the core has been removed. The x and y axes are normal and parallel to the arrow R, respectively. The white folded lines in Fig. 17b represent the profile of the tapered core buried in a gold metallic film. A part of the foot of the core was removed to form a sharp edge, where the height of the removed part was 1.5 µm. This probe was fabricated by the following four steps.

1. By the selective chemical etching technique, the GeO_2-doped core was tapered to $\alpha = 25°$ [5].
2. The foot of the core was removed by FIB.
3. The core was coated with a 500-nm-thick gold film.
4. The top of the core was removed to form an aperture with FIB.

Note that apertured probes with D as small as 30 nm has been realized with this fabrication process.

We compared the near-field optical intensity of a conventional single-tapered probe and that of two edged probes with $D = 500$ nm ($\sim D_c$ of EH_{11} mode) and $D = 100$ nm ($\sim \lambda/8$). Though both EH_{11} and HE-plasmon modes can propagate in a probe with $D = 500$ nm, note that the EH_{11} mode cannot exist in a probe with $D = 100$ nm.

First, we checked whether the edged structure of the probe led to efficient excitation of the HE-plasmon mode. We observed the spatial distribution of the optical near-field intensity in an edged probe with $D = 500$ nm. A protruding probe with an apex diameter (D_{apex}) of 10 nm and a foot diameter of 65 nm was used as a scanning probe (see Fig. 18) in the probe-to-probe method [11]. Figure 19a and b shows the observed distributions, where the

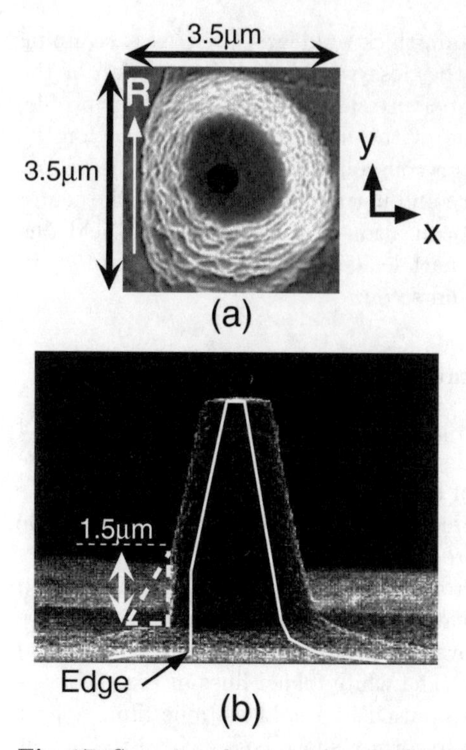

Fig. 17. Scanning electron micrographs of an edged probe. (**a**) Top view, (**b**) side view. The arrow R in (**a**) indicates the direction parallel to the surface from which the core was removed. The x and y axes are normal and parallel to the arrow R, respectively

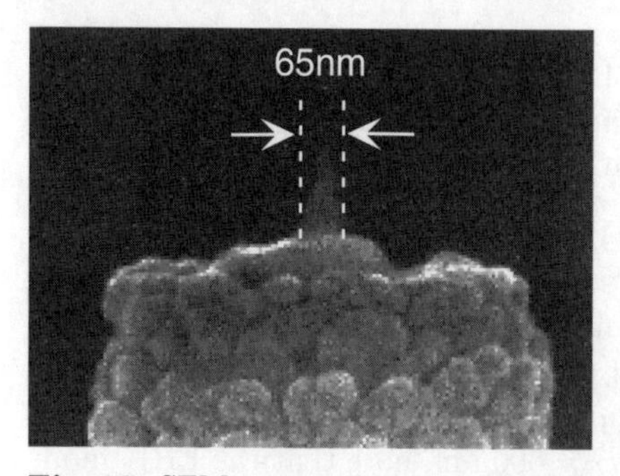

Fig. 18. SEM image of a protruding probe with $D_{\mathrm{apex}} = 10$ nm and foot diameter 65 nm

directions of incident light polarization were orthogonal to each other. Figure 19c and d shows calculated distributions, corresponding to the EH_{11} and HE-plasmon modes for $D = 500$ nm, respectively. Note that the distributions in Fig. 19a and b are in good agreement with those in Fig. 19c and d, re-

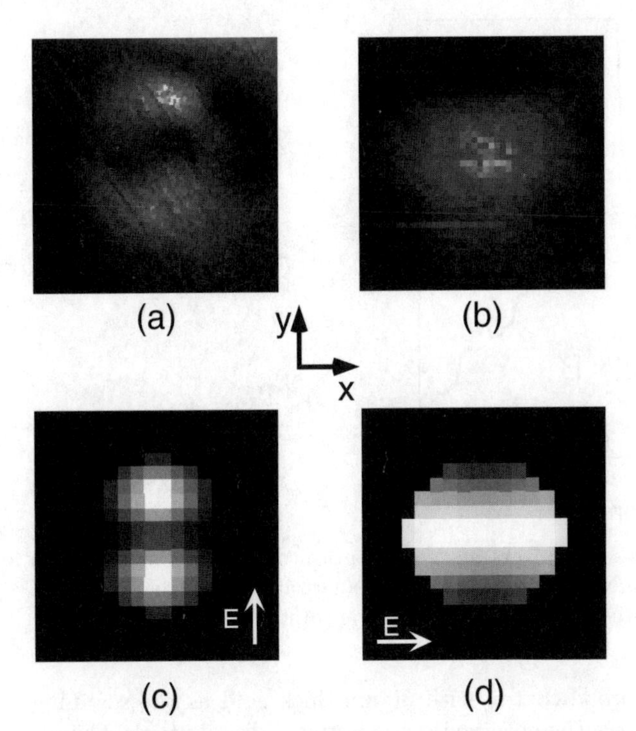

Fig. 19. Spatial distributions of the optical near field on top of an edged probe with $D = 500$ nm. The images are 750×750 nm. (**a, b**) Measured results. The directions of incident light polarization for these figures are orthogonal to each other. (**c, d**) Calculated results for the EH_{11} and the HE-plasmon mode, respectively. The vectors E in these figures represent the directions of polarization of the incident light

spectively, with respect to the number of lobes. Because the direction of light polarization changes in the fiber, we could not experimentally determine the direction of polarization at the foot of the probe. However, it seems most reasonable to consider the polarization directions of Fig. 19a and b to be those of Fig. 19c and d, respectively. We also observed the spatial distribution in a conventional probe (fabricated without step 2) with $D = 500$ nm. In this case, only a double-peaked distribution, which corresponds to the EH_{11} mode, was observed, where the lobes rotated as the direction of the incident light polarization changed. This indicates that an edged structure supports excitation of the HE-plasmon mode, where its excitation efficiency depends on the direction of incident light polarization.

Second, in order to check whether the optical near-field intensity in the apertured probe is enhanced due to the edged structure, we compared the spatial distributions of conventional and edged probes with $D = 100$ nm. To increase the efficiency of light scattering in the probe-to-probe method,

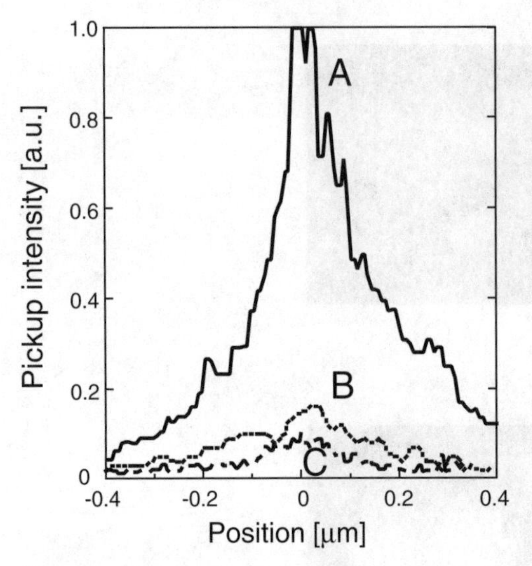

Fig. 20. Measured cross-sectional profiles of the optical near field. Curves A and B are for an edged probe, where the directions of polarization of the incident light are orthogonal to each other. Curve C is for a probe without an edge

we used a sharpened probe coated with 30-nm-thick gold as the scanning probe [8]. Figure 20 shows the observed cross-sectional distributions. Curves A and B are for the edged apertured probe, where the directions of incident light polarization were orthogonal to each other. A single-peaked distribution, which corresponds to the HE-plasmon mode, was observed. Curve C is for a conventional apertured probe. Note that the intensity for the curve A is ten times that of curves B and C. The FWHM of the intensity distribution of the HE-plasmon mode for $D = 100$ nm is analytically estimated to be 120 nm, which is comparable to that of curve A (~ 150 nm). This indicates that the increase in the intensity of the optical near field is a consequence of efficient excitation of the HE-plasmon mode at the sharp edge.

Finally, we compared the throughput of an edged probe, triple-tapered probe, and conventional probe. In Fig. 21, the throughput is plotted as a function of D. It shows that the throughput of the triple-tapered probe is 1000 times that of the conventional probe for $D < 100$ nm. Such a drastic increase in the throughput of the triple-tapered probe can be attributed to efficient excitation of an HE-plasmon mode. Since the dependence of the throughput on D for the triple-tapered probe is similar to that of an edged probe with 70 nm $< D <$ 200 nm (see Fig. 21), the HE-plasmon mode should be excited in the triple-tapered probe. As a consequence of the scattering coupling at the foot of the third taper [9], the HE-plasmon mode is excited efficiently. Furthermore, note that the triple-tapered probe has throughput 10 times that of the edged probe with $D < D_c$ (of the HE-plasmon mode). This

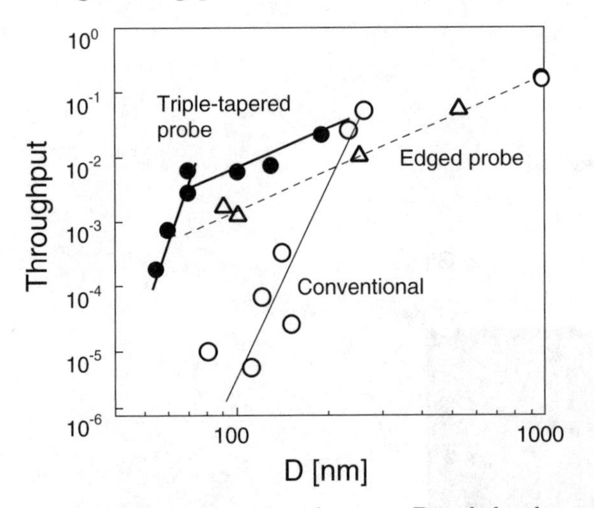

Fig. 21. Measured reations between D and the throughput

is because mode conversion from EH_{11} to HE-plasmon in the edged probe takes place at the foot of the core, where the guiding loss of the HE-plasmon mode is greater than that of EH_{11} mode. Thus, the edged probe has a greater guiding loss than a triple-tapered probe in which the mode is converted at the foot of the third-tapered core.

Pyramidal Silicon Probe

For further improvements in the performance of a near-field optical probe, we describe here the extremely high throughput and resolution capability of a pyramidal silicon probe. Since the high refractive index of the silicon ($n = 3.67$ at $\lambda = 830$ nm) leads to a short effective wavelength inside the probe, higher throughput and a smaller spot size are expected in comparison with conventional fiber probes [12].

A pyramidal silicon probe was fabricated from a (100)-oriented silicon on insulator (SOI) wafer by following four steps:

1. The SOI was bonded to the glass substrate by anodic bonding (300 V, 350°C, 10 min., Fig. 22a) [13].
2. After removing the silicon substrate, the SiO_2 layer was patterned by photolithography.
3. The single-tapered pyramidal probe was fabricated by anisotropic etching (40 g:KOH+60 g:H_2O+40 g:isopropyl alcohol at 80°C, Fig. 22b and c).
4. The probe was coated with 20-nm-thick aluminum (Fig. 22d).

Note that the pyramidal silicon probe with D_{apex} as short as 30 nm was realized by this process.

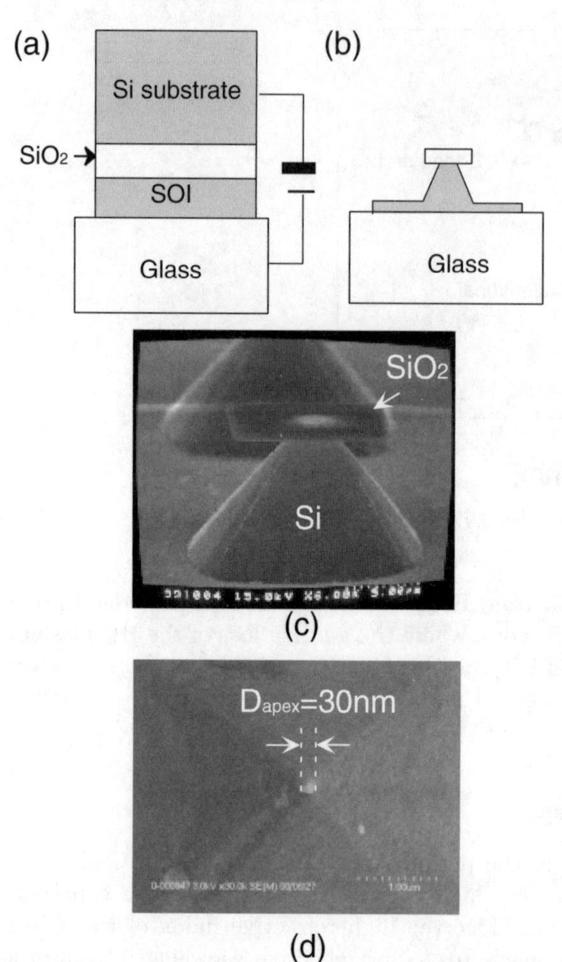

Fig. 22. (**a, b**) Schematic of the fabrication process of a pyramidal silicon probe. D_{apex}: apex diameter. SEM images of a pyramidal silicon probe: (**c**) bird's-eye view of the probe by step 3, (**d**) top view of the probe by step 4, respectively

We evaluated the throughput and spot size of a pyramidal silicon probe with $D_{\text{apex}} = 30$ nm. For comparison, we also evaluated a fiber probe with $D = 920$ nm ($= D_c$ of TE_{11} mode). As described in detail in Sect. 1.3, an apertured probe with $D = D_c$ (of TE_{11} mode) has a very small spot (FWHM=150 nm) due to the interference characteristics of the guided modes. The throughput has been calibrated to be 10 %. As a scanning probe for the probe-to-probe method, we used a fiber probe with D of 50 nm.

Figure 23a and b shows the observed spatial distribution of the optical near-field intensity for a pyramidal silicon probe with $D_{\text{apex}} = 30$ nm and an apertured fiber probe with $D = 920$ nm, respectively. Curves A and B in

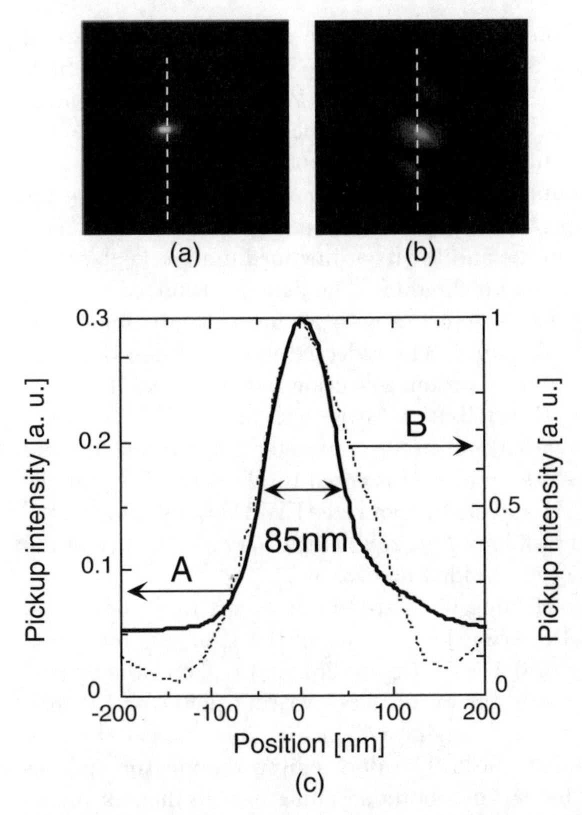

Fig. 23. Spatial distribution of the optical near-field intensity. (**a**) and (**b**) are for a pyramidal silicon probe with $D_{\text{apex}} = 30$ nm and a fiber probe with $D = 920$ nm, respectively. The images are 1.5×1.5 µm. Curves A (*solid line*) and B (*dashed line*) in (**c**) are the cross-sectional profiles along the dashed white line in (**a**) and (**b**), respectively

Fig. 23c show the cross-sectional profiles along the dashed white line in (a) and (b), respectively. Note that the FWHM of curve A is as narrow as 85 nm ($\sim \lambda/10$). Furthermore, the peak intensity of curve A is 25 % that of B. This indicates that a pyramidal silicon structure can yield high throughput and small spot size simultaneously.

2 Application to High-Density and High-Speed Optical Memory

2.1 Using an Apertured Fiber Probe

We applied a high-throughput fiber probe for optical recording and reading. An as-deposited phase-change material in an amorphous phase was used as the recording medium. The phase-change film on a glass substrate consisted of a 20-nm-thick SiO_2 film, a 25-nm-thick $AgInTe_2$, and a 20-nm-thick SiO_2 film. These films were deposited by RF sputtering. We carried out the phase-change recording and reading with the high-throughput fiber probe. The experimental setup is shown in Fig. 24. Linearly polarized light from a laser diode ($\lambda = 830$ nm) was coupled into the single-mode fiber to generate the optical near field on the aperture. The irradiation time was fixed to 2 μs. The separation between the probe and the recording medium was kept within several nanometers by shear-force techniques. The light transmitted through the aperture and the sample was detected with a photodetector under the medium (illumination mode; I mode). The reflected light was also detected with a photomultiplier tube (illumination–collection mode; I–C mode).

Figure 25 shows the spatial distribution of the optical near-field intensity on the fiber probe, with $D = 920$ nm used for phase-change recording and reading. As described in Sect. 1.3, since D is equal to D_c of the TE_{12} mode, it has high throughput (10 %) and small spot size (FWHM = 180 nm) at the center of the cross-sectional profile in Fig. 25b, which is a consequence of the interference characteristics of the guided modes.

After optical near-field recording was carried out over a range of optical recording power, the spatial distribution of transmitted light was measured at an optical reading power of 0.2 mW. Figure 26a and b shows the spatial distribution obtained by I mode operation for a writing power of 7.0 mW and 8.6 mW, respectively. A small bright spot in Fig. 26b (indicated by an arrow) represents the recorded mark. The dark ellipse around the spot as well as that in Fig. 26a is due to the photodarkening in thin films of amorphous chalcogenides [14,15]. Since Te, being a two-fold coordinated atom, has chain-molecule fragments [16], structural changes are induced by exposure light. This results in an increase in the absorption coefficient, which is called photodarkening [16]. The photodarkening might be caused by the side lobes of the curve in Fig. 25b.

Figure 26c shows the spatial distributions obtained by I-C mode operation for a writing power of 8.6 mW. Note the dark small spot at the center and a ring, which are due to photodarkening. In Fig. 26d, curves A and B are the cross-sectional profiles along the dashed white lines in (b) and (c), respectively. Note that the FWHM of these curves is as narrow as 250 nm.

2.2 High-Density and High-Speed Recording Using a Pyramidal Silicon Probe on a Contact Slider

To realize a high data transmission rate, a superresolution near-field structure [18] and a planar probe mounted on an optical slider [19] have been proposed recently. Since these methods do not require shear-force feedback, reading speed is increased without technical difficulties. However, since the incident light has to be focused by a lens in these systems, the efficiency of the optical near-field generation on a subwavelength aperture is low.

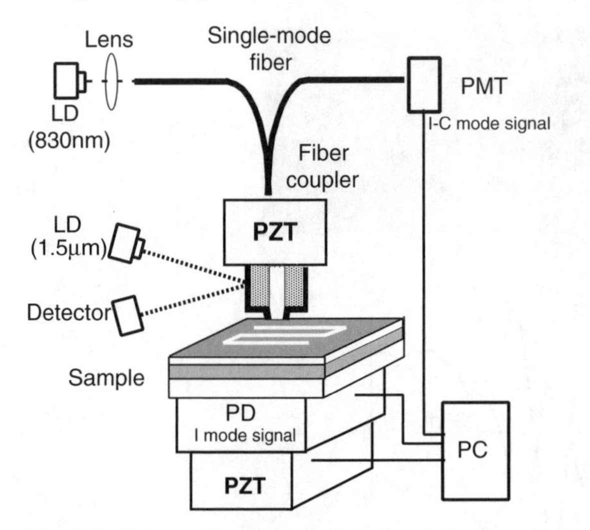

Fig. 24. Schematic of the system used for near-field phase-change recording and reading. PD, photodetector for I mode signal; PMT, photomultiplier tube for I-C mode signal

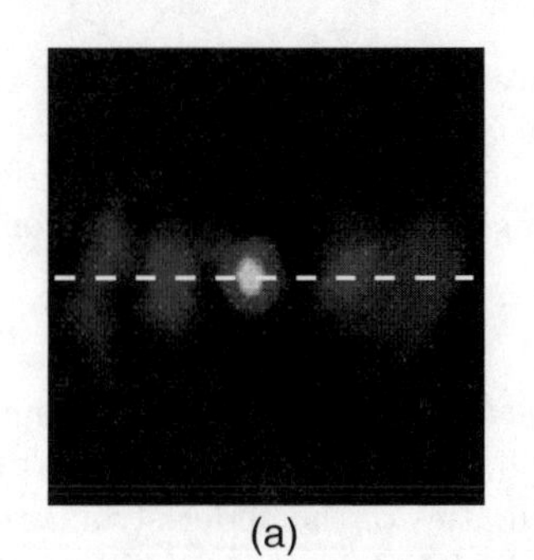

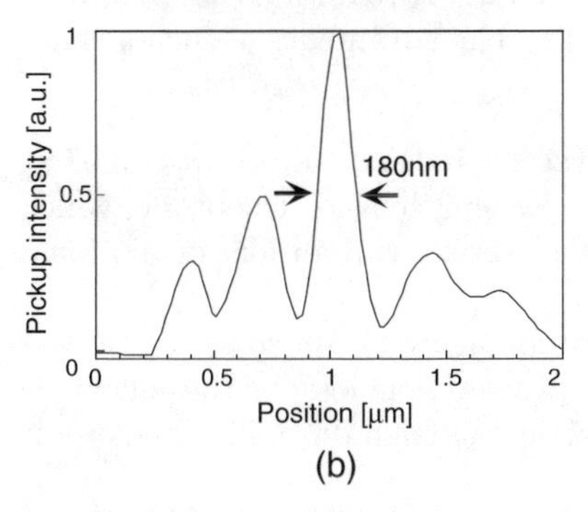

Fig. 25. Spatial distribution of the optical near-field intensity on the aperture, with $D = 920$ nm used for phase-change recording. (**a**) Two-dimensional image. (**b**) Cross-sectional image along the dashed white line in (**a**)

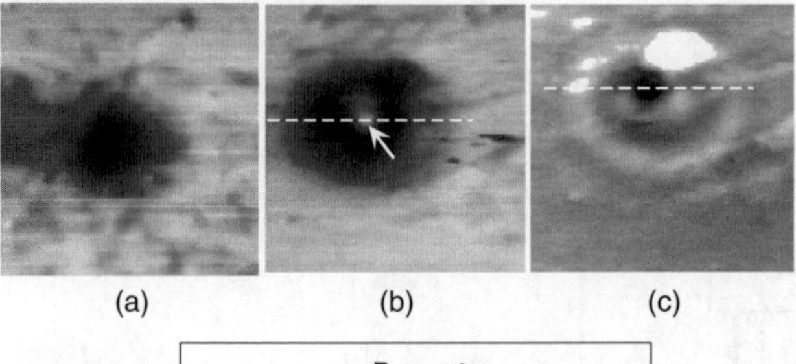

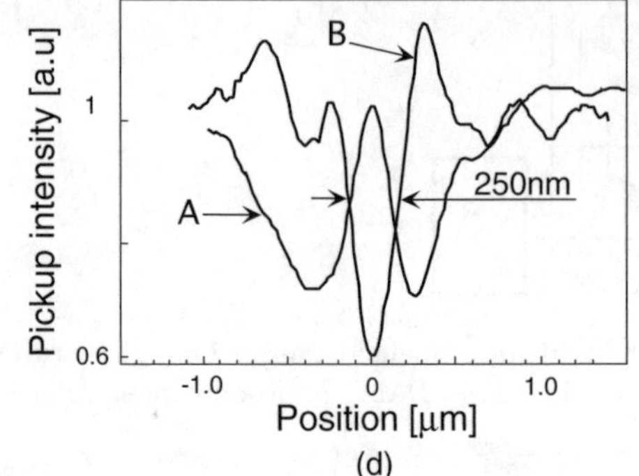

Fig. 26. Spatial distribution of the optical near-field intensity images of phase-change recordings. (**a**) and (**b**) obtained by I mode operation for writing power of 7 mW and 8.6 mW, respectively. (**c**) obtained by I-C mode operation for writng power of 8.6 mW. (**d**) Curves A and B show the cross-sectional profiles along the dashed white lines in (**b**) and (**c**), respectively

In this section, we describe a new contact slider with a high throughput of optical near-field intensity to realize a high recording density on the phase-change medium and fast readout [17]. Schematics of the slider structure and the data storage system are illustrated in Fig. 27. A pyramidal silicon probe array is arranged at the rear end of the slider. The advantages of such a slider are:

1. As described in Sect. 1.4, the high refractive index of silicon ($n = 3.67$ at $\lambda = 830$ nm) leads to a short effective wavelength inside the probe, which results in higher throughput and smaller spot size than for conventional fiber probes made of silica glass.
2. The height of the pyramidal silicon probe array is controlled to be less than 10 μm so that sufficiently low propagation loss in the silicon is maintained. Furthermore, the probe array has high durability because it is bonded to a thick glass substrate.
3. Compared with those of previously reported pyramidal probes fabricated by use of a focused ion beam [20] or by the transfer mold technique in a pyramidal silicon groove [21], ultrahigh homogeneity in the heights of the probes and pads can be realized. This is because the flatness of the mesa

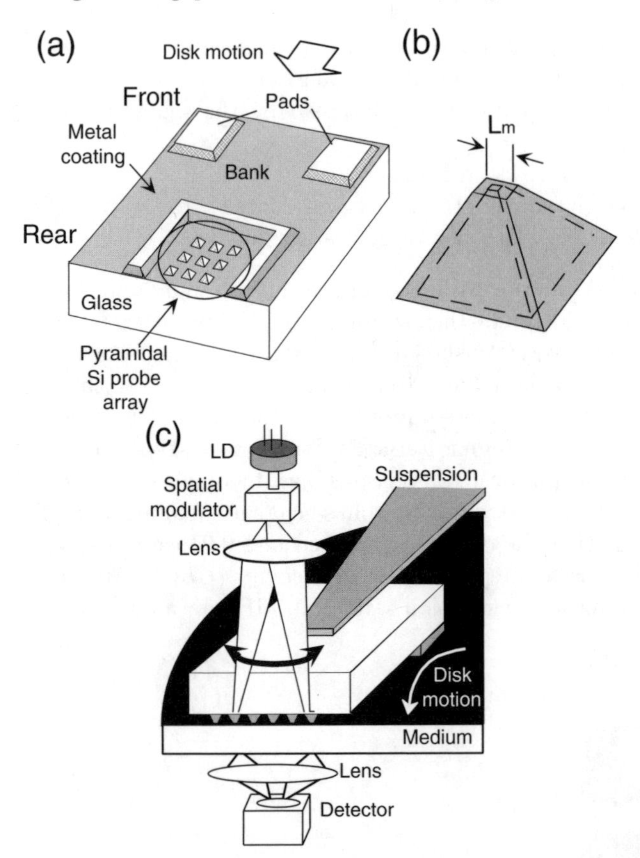

Fig. 27. (**a**) Contact slider with a pyramidal silicon probe array. (**b**) Pyramidal silicon probe with a mesa. L_m is the mesa length. (**c**) Schematic of the storage system with the slider: LD, laser diode

at the probe tip and of the upper surface of the pads are determined by the uniformity of the thickness of silicon wafer.

4. Use of a probe array with many elements increases the total data transmission rate by parallel readout [21–23]. In this system the incident light is spatially modulated by an electro-optical modulator, and the scattered light from a different probe can be read out as a time-sequential signal.

Since the key issue in realizing a pyramidal silicon probe array is high homogeneity in the heights of the probes, the probe array is fabricated from a (100)-oriented SOI wafer in four steps:

1. The SOI wafer was bonded to the glass substrate by anodic bonding (300 V, 350°C, 10 min.) [13].
2. After removing the silicon substrate from the SOI wafer by mean of wet etching, the SiO_2 layer was patterned by photolithography.

3. The probe array, the bank, and the pads were fabricated by means of anisotropic etching (40 g:KOH+60 g:H_2O+40 g:isopropyl alcohol, 80°C). Note that height homogeneity was maintained by the remaining SiO_2 layer.
4. The slider was quarried with a dicing saw.
5. After removing the SiO_2 layer, the slider was coated with 30-nm-thick aluminum to increase the efficiency of light scattering [8].

Figure 28a and b shows an optical image of the contact slider and a scanning electron microscopic image of the pyramidal silicon probe array fabricated on the slider, respectively. As shown in Fig. 28b, the height dispersion of the probes and pads has been reduced to less than 10 nm, because this dispersion is determined by the thickness uniformity of the SOI wafer. Here the slider is designed by use of the design criteria [24] for a contact-type hard-disk head so that its jumping height over the phase-change medium is maintained to less than 10 nm. Furthermore, since the phase-change medium is fragile, we designed the bank so that the contact stress becomes 0.01 times the yield stress of the magnetic disk at a constant linear velocity (*CLV*) of 0.3 m/s, which corresponds to a data transmission rate of 10 MHz for a data storage

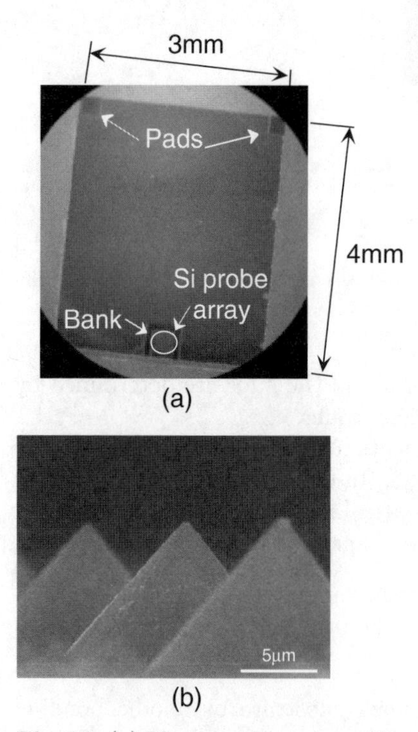

Fig. 28. (**a**) Photographic image of the contact slider with a pyramidal silicon probe array and pads. (**b**) Magnified scanning electron microscope image of the fabricated pyramidal silicon probe array

density of 1 Tbit/in^2. To increase the readout speed 100-fold, i.e., to realize a 1 Gbit/s data transmission rate for data storage density of 1 Tbit/in.^{2}, we fabricated 100 probe elements on the inner part of the bank for parallel readout.

In recording and readout experiments with the fabricated contact slider, we compared the signal transmitted through phase-change marks recorded with a single element of the probe array and that recorded with focused propagating light. Figure 29 shows the experimental setup. The contact slider was glued to a carrier. The slider was in contact with a phase-change medium coated with a thin lubricant film (Fomblin Z-DOL). A laser beam ($\lambda = 830$ nm) was focused on one element of the probe array on the slider. The frequency of the rectangularly modulated signal with 50 % duty cycle was varied from 0.16 to 2.0 MHz at a $CLV = 0.43$ m/s. The light transmitted through the recording medium was detected with an avalanche photodiode. We used an as-deposited AgInSbTe film as a recording medium. The optical recording powers for a pyramidal silicon probe with a mesa length L_m of 150 nm (see Fig. 30a) and propagating light focused by a microscope objective ($N.A. = 0.40$) were 200 mW and 15 mW, respectively, which are, to the authors' knowledge, the lowest recording powers. The throughput of optical near-field generation of the metallized pyramidal silicon probe is 7.5×10^{-2}, which is estimated from the ratio of the optical powers for near- and far-field recording. Readout was carried out at a $CLV = 0.43$ m/s. The reading optical powers for the pyramidal silicon probe and the focused propagating light were 20 mW and 3.6 mW, respectively. The resolution bandwidth was fixed at 30 kHz.

The dependence of the carrier-to-noise ratio (CNR) on the mark length is shown in Fig. 30b. Note that shorter crystalline phase-changed marks beyond the diffraction limit were recorded and read out by an optical near field

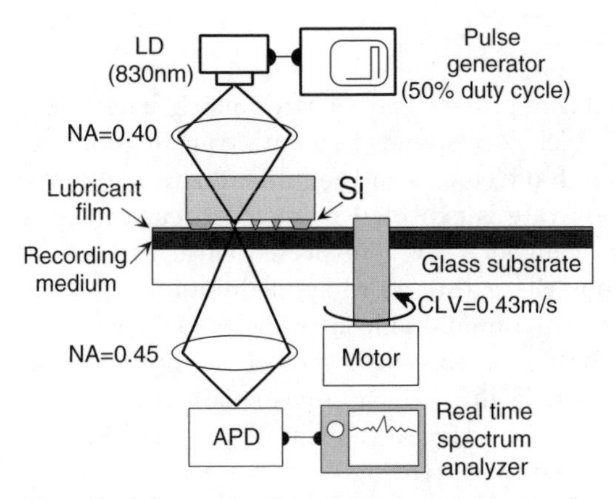

Fig. 29. Schematic of the experimental setup for phase-change recording and reading by the contact slider. APD, avalanche photodiode; LD, laser diode; NA, numerical aperture

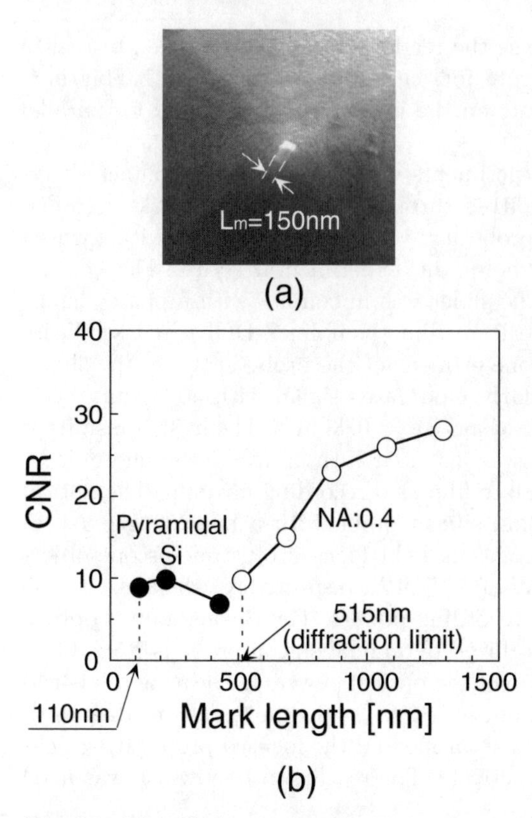

Fig. 30. Magnified scanning electron microscope image of the pyramidal silicon probe tip used for optical near-field recording and reading. L_m, mesa length. (**b**) *CNR* dependence on the mark length for (*closed circles*) the optical near field with the pyramidal silicon probe and (*open circles*) propagating light focused by an objective

generated on the pyramidal silicon probe. The shortest mark length was 110 nm at a *CLV* of 0.43 m/s, which corresponds to a data transmission rate of 2.0 MHz. Since this slider has 100 elements in the probe array, a 100-fold increase in the data transmission rate is expected by parallel readout using all the elements. Furthermore, a higher *CLV* is expected since we did not observe any damage on the probe tip or the recording medium after a series of experiments. The *CNR* of the pyramidal silicon probe seen in Fig. 30b was indepencent of mark length, due to the small spot of the optical near-field and the narrow recorded mark width, which are as small as L_m of the pyramidal silicon probe. These results indicate that an increased *CNR* and shorter mark length can be achieved by employing servo-control for tracking during readout. Furthermore, it is expected that the recording density can be increased to as high as 1 Tbit/in^2 by optimization of the interference characteristics of the guided modes in the pyramidal silicon probe.

3 Outlook

Higher throughput is expected by improving the techniques reviewed in this chapter. In addition to these techniques, use of the plasmon effect is advantageous for further increases in throughput [25,26]. For extension to the ultraviolet, a pure silica core fiber of low transmission loss has been developed and used to fabricate a high-throughput fiber probe [27]. For improvements in mechanical stability, micro-machining technology is advantageous for fabricating a silicon probe. A silicon probe with throughput as high as 2.5 % (see Sect. 1.4) and a hollow silicon probe with 20 nm aperture diameter for the ultraviolet [28] have been fabricated using this technique. Furthermore, functional probes, e.g., probes on which light-emitting dye molecules [29] and ZnO films [30] are deposited, are attractive for their optical frequency conversion capabilities which can be used in biological applications.

Section 2 was devoted to describing the application of high-throughput probes to optical memory. However, one can find further applications. These are spatially resolved spectroscopy in the ultraviolet, visible, and infrared by detecting photoluminescence, electroluminescence, photocurrent, and Raman signals of semiconductor quantum dots [31], semiconductor devices [32], and organic materials [33]. Application to nano-fabrication, e.g., photochemical vapor deposition of nano-structures of metals (Zn and Al), oxides (ZnO), and composite semiconductors has also been demonstrated in order to realize high-density nano-photonic integrated circuits for future optical transmission systems [34–36]. These techniques can open a new field of nano-photonics. Further, high-throughput fiber probes can be also used to manipulate neutral atoms in vacuum by controlling their thermal motion, which can open a new field of atom-photonics [36].

Acknowledgements

Valuable discussions with Drs. M. Kourogi, K. Tsutsui (Tokyo Institute of Technology), J. Tominaga (National Institute of Advanced Industrial Science and Technology), and J. Takahashi (Ricoh Company, Ltd.) are deeply appreciated.

References

1. M. Ohtsu (ed.), *Near-Field Nano/Atom Optics and Technology* (Springer, Tokyo 1998)
2. L. Novotny, C. Hafner: Phys. Rev. E **50**, 4094 (1994)
3. E.D. Palik (ed.), *Handbook of Optical Constants of Solids* (Academic Press, San Diego 1985)
4. T. Yatsui, M. Kourogi, M. Ohtsu: Appl. Phys. Lett. **71**, 1756 (1997)
5. T. Pangaribun, K. Yamada, S. Jiang, H. Ohsawa, M. Ohtsu: Jpn. J. Appl. Phys. **31**, L1302 (1992)

6. G.A. Valaskovic, M. Holton, G.H. Morrison: Appl. Opt. **34**, 1215 (1995)
7. T. Yatsui, M. Kourogi, M. Ohtsu: Appl. Phys. Lett. **73**, 2090 (1998)
8. Y. Inoue, S. Kawata: Opt. Lett. **19**, 159 (1994)
9. D. Marcuse: *Light Transmission Optics*, 2nd edn. Chap. 10 (Krieger, Malabar 1982)
10. A. Sommerfeld: *Optics*, Chap. VI, 39 (Academic Press, New York 1954)
11. S. Mononobe, M. Naya, T. Saiki, M. Ohtsu: Appl. Opt. **36**, 1496 (1997)
12. H.U. Danzebrink, T. Dziomba, T. Sulzbach, O. Ohlsson, C. Lehrer, L. Frey: J. Microscopy **194**, 335 (1999)
13. T.R. Anthony: J. Appl. Phys. **58**, 1240 (1998)
14. F. Yonezawa (ed.), *Fundamental Physics of Amorphous Semiconductors*,104 (Springer, Berlin Heidelberg New York 1981)
15. K. Tanaka: J. Non-Crys. Solids **59** & **60**, 925 (1993)
16. B.W. Corb, W.D. Wei, B.L. Averbach: J. Non-Crys. Solids **53**, 29 (1994)
17. T. Yatsui, M. Kourogi, K. Tsutsui, J. Takahashi, M. Ohtsu: Opt. Lett. **25**, 1279 (2000)
18. J. Tominaga, T. Nakano, N. Atoda: Appl. Phys. Lett. **73**, 2078 (1998)
19. H. Yoshikawa, Y. Andoh, M. Yamamoto, K. Fukuzawa, T. Tamamura, T. Ohkubo: Opt. Lett. **25**, 67 (2000)
20. F. Isshiki, K. Ito, S. Hosaka: Appl. Phys. Lett. **76**, 804 (2000)
21. Y.J. Kim, K. Kurihara, K. Suzuki, M. Nomura, S. Mitsugi, M. Chiba, K. Goto: Jpn. J. Appl. Phys. **39**, 1538 (2000)
22. T. Yatsui, M. Kourogi, K. Tsutsui, J. Takahashi, M. Ohtsu: In: *Proc. SPIE 3791* (1999) p. 76
23. D.W. Pohl: IBM J. Res. Dev. **39**, 701 (1995)
24. M. Yanagisawa, A. Sato, K. Ajiki, F. Watanabe: In: *Tribology of Contact/Near-Contact Recording for Ultra High Density Magnetic Storage.* ed. by C.S. Bhatia, A.K. Menon, TRIB-**6** (American Society of Mechanical Engineers, New York, 1996) p. 25
25. M. Ashino, M. Ohtsu: Appl. Phys. Lett. **72**, 1299 (1998)
26. T. Matsumoto, T. Ichimura, T. Yatsui, M. Kourogi, T. Saiki, M. Ohtsu: Opt. Rev. **5**, 369 (1998)
27. S. Mononobe, T. Saiki, T. Suzuki, S. Koshihara, M. Ohtsu: Opt. Commun. **146**, 45 (1998)
28. T. Yatsui, M. Kourogi, M. Ohtsu: In: *Technical Digest of Quantum Electronics and Laser Science Conference* (2001) p. 79
29. K. Kurihara, M. Ohtsu, T. Yoshida, T. Abe, H. Hisamoto, K. Suzuki: Anal. Chem. **71**, 3558 (1999)
30. Y. Yamamoto, G.-H. Lee, K. Matsuda, T. Shimizu, M. Kourogi, M. Ohtsu: Opt. Rev. **7**, 486 (2000)
31. T. Saiki, K. Nishi, M. Ohtsu: Jpn. J. Appl. Phys. **37**, 1638 (1998)
32. H. Fukuda, Y. Kadota, M. Ohtsu: Jpn. J. Appl. Phys. **38**, L571 (1999)
33. Y. Narita, T. Tadokoro, T. Ikeda, T. Saiki, S. Mononobe, M. Ohtsu: Appl. Spectro. **52**, 1141 (1998)
34. V.V. Polonski, Y. Yamamoto, M. Kourogi, H. Fukuda, M. Ohtsu: J. Microscopy **194**, 545 (1999)
35. Y. Yamamoto, M. Kourogi, M. Ohtsu, V. Polonski, G.-H. Lee: Appl. Phys. Lett. **76**, 2173 (2000)
36. M. Ohtsu, K. Kobayashi, H. Ito, G.-H. Lee: Proc. IEEE **88**, 1499 (2000)

Modulation of an Electron Beam in Optical Near-Fields

J. Bae, R. Ishikawa, and K. Mizuno

1 Introduction

Many kinds of electron beam devices have been developed and utilized in various scientific areas since the early 1900's [1]. For instance, microwave amplifiers and oscillators, electron accelerators, and various types of electron microscopes have made major contributions to establish modern science and technology, such as high-speed communication, elementary particle physics, and solid-state physics. On the other hand, most of those for commercial use have been replaced by solid-state devices due to their power consumption and device size. Beam devices, however, have several advantages over solid-state devices, such as wider tuning frequency range and higher output power. Recent advances in vacuum microelectronics [2] and micromachining technology [3] provide a useful means for realizing small beam devices. In fact, electron emitters with dimensions of several microns have been fabricated through semiconductor processes [4].

The operating frequencies of conventional beam devices are usually less than 1 THz [5]. Though free-electron lasers can operate at optical frequencies, these utilize relativistic electron beams for laser action, and consequently their device size is huge [6]. Lower-energy electron-beam devices operating at optical frequencies would thus be thus desirable development.

In order to realize compact, convenient beam devices using nonrelativistic electrons, the interaction mechanism for energy and momentum exchange between electron and light, including quantum effects, must be carefully studied. Those studies are also required to develop new microscopic techniques that are able to investigate the surface state of nanoparticles. For instance, electron energy loss spectroscopy (EELS) [7] is one of those, in which electrons passing in the proximity of a nanoparticle excite near-fields on its surface and suffer specific energy loss resulting from the interaction with the fields. Analyses for the energy-loss spectrum of the electrons bring out useful information about the electronic and optical properties of the nanoparticle [8].

In this chapter, energy modulation of nonrelativistic electrons with light waves, especially with evanescent waves contained in optical near-fields, are discussed. We shall briefly review in the next section electron–light interaction effects, called the Smith–Purcell effect and the Schwarz–Hora effect. In Sect. 3, a basic principle of electron–light interaction is explained in both classical and quantum terms. In Sect. 4, microgap circuits used for the interaction are presented. In Sect. 5, a metal microslit among those circuits is chosen as a potential circuit to realize the interaction with light waves. The energy modulation of low-energy electrons with optical evanescent waves in a microslit circuit have been demonstrated experimentally. The results are presented in Sect. 6. In Sect. 7, a multiple-slit circuit structure, this is, a diffraction grating, is treated as an interaction circuit. Based on the results, we discuss in Sect. 8 the possibility of experimental observation of electron-energy modulation with visible light. Section 9 concludes this chapter.

2 Review of Experiments

Low-energy electron-beam amplifiers and oscillators, klystrons, traveling wave tubes, and backward wave oscillators, are mostly for the microwave region. For these devices, it has been thought that extending their operation to optical frequencies would be difficult. The reason for this is as follows. In order to obtain signal gain in the beam devices, the electron beam must be density-modulated at a signal frequency. In the microwave region, density modulation (bunching) is easily achieved, because the energy of the electrons changes in proportion to the amplitude of the signal electric field. On the other hand, in the optical region, the electron energy changes by multiples of the energy of a signal photon. This effect would prevent a smooth change in beam density, diminishing the signal gain. In 1954, Senitzky concluded through his theoretical consideration that operation of klystron-type beam devices is limited to frequencies below submillimeter waves [9]. Contrary to his theoretical prediction, light emission effects with a nonrelativistic electron beam have been reported. In this section, two well-known effects, the Smith–Purcell effect and the Schwarz–Hora effect, are reviewed.

2.1 Smith–Purcell Effect

In 1953, Smith and Purcell demonstrated that light was emitted from an electron beam passing close to the surface of a metal diffraction grating [10]. Figure 1 shows the experimental configuration of the Smith–Purcell effect. They observed that the wavelength of the light could be changed from 450 nm to 550 nm by adjusting the initial energy of electrons from 340 keV to 309 keV. The relationship between the wavelength λ and the electron velocity v is given by

$$\lambda = \frac{D}{n}\left(\frac{1}{\beta} - \cos\theta\right) , \qquad (1)$$

where D is the period of the grating, n is the spectral order, β is the electron velocity normalized to the light speed c, i.e., v/c, and θ is the angle between the direction of motion of the electron and the light ray. This equation was

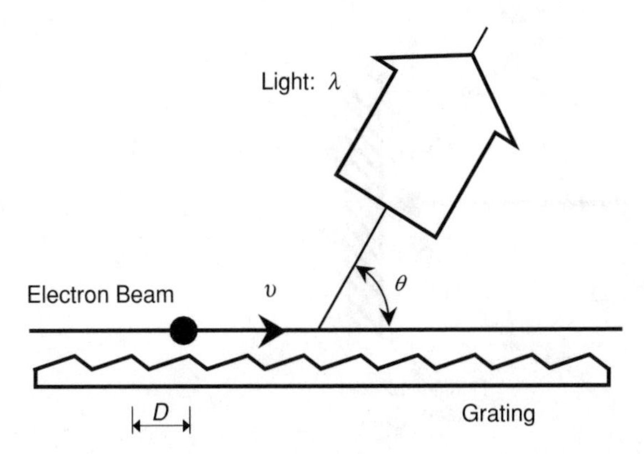

Fig. 1. Schematic drawing of the Smith–Purcell effect

derived by them through a simple Huygens analysis for radiation caused by the periodic motion of the charge induced on the surface of the grating. This emission of radiation from the electrons has been also explained in terms of the interaction between the electrons and evanescent surface waves induced on the grating [11].

Smith and Purcell used relatively high energy electrons with $\beta \sim 0.8$ for their experiment. In 1997, Goldstein et al. successfully observed Smith–Purcell radiation using a lower-energy electron beam with an energy of 27 keV to 40 keV, though the radiation was in the far infrared [12]. After the report by Smith and Purcell, many researchers proposed various types of beam devices using the Smith–Purcell effect and its inverse [13], such as the laser-driven grating linac [4,5] and free electron lasers with a grating [16]. Most of those devices have not yet been demonstrated experimentally in the optical region.

2.2 Schwarz–Hora Effect

In 1969, Schwarz and Hora reported that quantum modulation of nonrelativistic electrons with light had been achieved using a thin dielectric film as an interaction circuit [17]. Figure 2 shows a schematic drawing of their experimental system. In the experiment, a 50-keV electron beam and an Ar laser (wavelength: $\lambda = 488$ nm) were used. The interaction circuit was a SiO_2 or Al_3O_2 film with a thickness of less than 200 nm. The electron beam passed through the film illuminated with the 10-W laser beam and then hit a nonluminescent screen. They reported that light of the same color as the laser was emitted from the screen. For the experimental parameters used by Schwarz and Hora, the laser produces a maximum electric field of about 10^7 V/m in the dielectric film. This field only gives an energy of 2 eV or less to the electrons in the film. This energy is less than the photon energy, 2.54 eV of the Ar laser, so that quantum effects might come into play in this interaction.

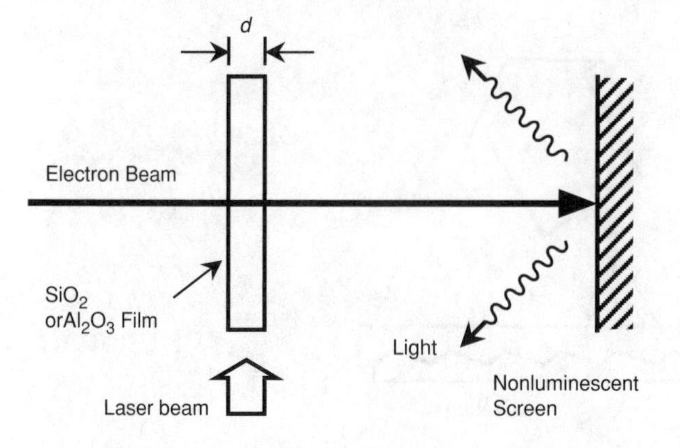

Fig. 2. Schematic drawing of the Schwarz–Hora effect

They claimed that the wave function of coherent electrons was modulated by the light, and consequently density modulation of the electron beam was achieved [18]. Though the Schwarz–Hora effect has not yet been reproduced by other experimental groups, their results have suggested the possibility of new electron beam devices utilizing quantum effects.

3 Basic Principle

Though the Schwarz–Hora effect has not been confirmed experimentally yet, we believe that a thin dielectric film could be used as an interaction circuit between electrons and light. This is because energy and momentum conservation are satisfied by the interaction [19]. Let us consider the case in which an electron absorbs a photon. The dispersion relations for the electron and photon are represented by (2) and (3), respectively, and are illustrated in Fig. 3:

$$W_e = \sqrt{{m_0}^2 c^4 + c^2 {p_e}^2}\,, \tag{2}$$

$$W_p = \hbar\omega = c p_p\,, \tag{3}$$

where W and p indicate the energy and the momentum, respectively, and the subscripts e and p stand for electron and photon, m_0 is the rest mass of the electron, c is the speed of light, $\hbar = h/2\pi$, h is Planck's constant, and ω is the angular frequency. In the thin dielectric film shown in Fig. 2, p_e is perpendicular to p_p. Therefore, the increase in electron momentum p_e when W_e increases by W_p is

$$\Delta p = \frac{\hbar\omega}{v}\,, \tag{4}$$

where v is the electron velocity. The interaction between an electron and a photon in free space does not occur due to this shortage of momentum.

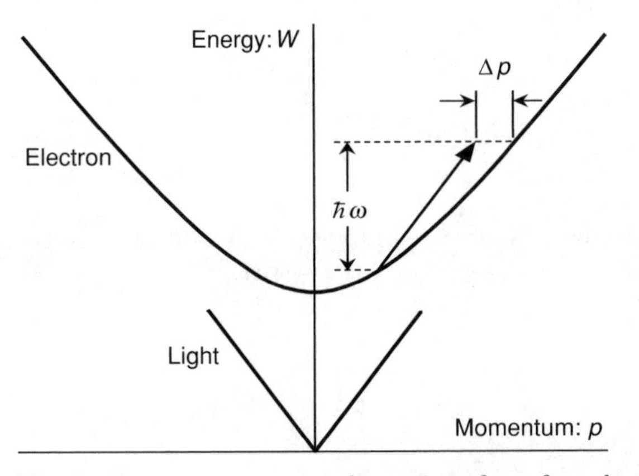

Fig. 3. Energy–momentum dispersions for a free electron and a photon

From an evanescent wave theory [20], it is known that localized optical fields in an area small compared to the wavelength, such as optical near-fields, have wave components whose momentum is greater than that of free photons. Therefore, those fields could modulate an electron beam. This argument is supported by the experimental results in the Smith–Purcell effect described in Sect. 2.1, in which electrons interact with optical evanescent waves induced on the surface of a metal grating, and coupling to the field gives rise to radiation.

For ease of consideration, let us assume a simple object: a slit of infinite length in the z direction, as shown in Fig. 4. When the slit is illuminated by light with angular frequency ω, an optical near-field is induced on the surface of the slit. The electric field E_x on the slit can be expressed in terms of its spatial plane-wave spectrum A, and is given by

$$E_x(x,t) = \frac{1}{2\pi} \int_{-\infty}^{+\infty} A(k_x) \exp j(k_x x - \omega t) dk_x \, , \tag{5}$$

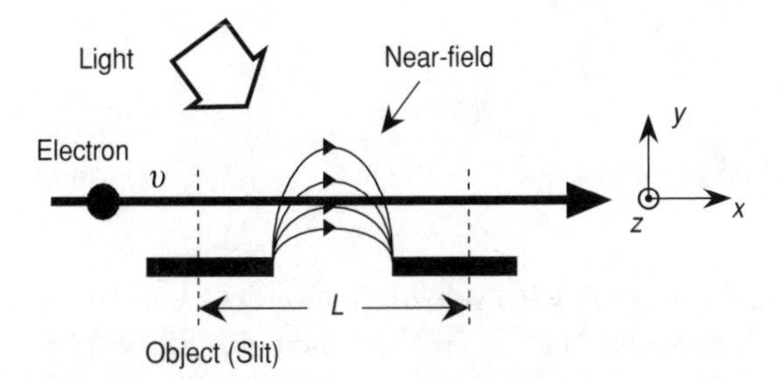

Fig. 4. Theoretical model for energy changes of electrons moving with a velocity v through an optical near-filed region on an small object (slit)

where

$$A(k_x) = \int_{-\infty}^{+\infty} E_x(x,t) \exp[-j(k_x x - \omega t)] dx \; . \tag{6}$$

When an electron passes through the near-field region with velocity v in the x direction, the electron suffers an energy loss (or gain) W,

$$W = \int_{-L/2}^{+L/2} q E_x(x,t) dx \; , \tag{7}$$

where L is the interaction length on the slit and q is the electron charge. Substituting (5) for E_x into (7), we obtain

$$W = \frac{q}{2\pi} \int_{-L/2}^{+L/2} \int_{-\infty}^{+\infty} A(k_x) \exp j(k_x x - \omega t) dk_x dx \; . \tag{8}$$

Using the relation $t = x/v$, it follows that

$$W = q \int_{-\infty}^{+\infty} A(k_x) \left\{ \frac{L}{2\pi} \frac{\sin(k_x - k_e)\frac{L}{2}}{(k_x - k_e)\frac{L}{2}} \right\} dk_x \; , \tag{9}$$

where

$$k_e = \omega / v \; . \tag{10}$$

The sampling function in (9) becomes a delta function in the limit $L \to \infty$:

$$\lim_{L \to \infty} \left\{ \frac{L}{2\pi} \frac{\sin(k_x - k_e)\frac{L}{2}}{(k_x - k_e)\frac{L}{2}} \right\} = \delta(k_x - k_e) \; .$$

Hence we have

$$W = q \int_{-\infty}^{+\infty} A(k_x) \delta(k_x - k_e) dk_x = q A(k_e) \; . \tag{11}$$

From this result, it is seen that an electron with velocity v interacts selectively with the wave component with wave number k_e in the near-field. Since k_e is greater than the wave number $k_0 = \omega / c$ of light in free space, this wave is evanescent. Since the momentum of the evanescent wave, $\hbar k_e$, is greater than that of a wave propagating in free space, energy and momentum conservation are satisfied for the interaction with electrons. In fact, substituting $\hbar k_e$ for Δp in (4), the same expression, i.e., (10), can be derived.

An analogous treatment of the interaction has been carried out using the uncertainty principle. When optical fields are localized on a slit with a small width d, the effective interaction length on the slit is limited, so that an uncertainty exists in momentum. This additional momentum p_c from the slit is approximately

$$p_c \sim \frac{h}{d} . \tag{12}$$

When $p_c > \Delta p$, momentum conservation for the interaction can be also satisfied. From (4) and (12), the condition for d is

$$d < \beta\lambda . \tag{13}$$

This equation indicates that the requirement for interaction between electrons and light is not the slit itself, but limitation of the interaction length. From (13), it is seen that for Schwarz and Hora's experiment, significant interaction can occur when the film thickness is less than 200 nm.

4 Microgap Interaction Circuits

Three kinds of circuit structures for the interaction between free electrons and light are presented. The transition rates of electrons in the interaction circuits are theoretically estimated in quantum mechanical fashion. The results indicate that measurable signal electrons can be obtained under practical experimental conditions.

4.1 Circuit Configuration

As discussed in Sect. 3, localized field distributions are necessary for interaction with electrons. Figure 5 shows three kinds of circuits used in the optical region: (a) a dielectric film, (b) a metal film gap, and (c) a metal slit. These microgap circuits localize optical fields in a small gap of width d less than the wavelength.

In Fig. 5, the dielectric film is the interaction circuit used by Schwarz and Hora. The metal film gap is most similar to the conventional microwave circuits used in klystrons [5]. The metal microslit was proposed as an optical near-field generator used for the interaction with low-energy electrons [21]. This type of metal microslit is more suitable than the other gap circuits for measurement of energy exchange between free electrons and light and for investigating quantum effects, because there is not a disturbance such as electron scattering in metal films or dielectric medium. On the other hand, the near-fields only exist in proximity to the slit surface so that the interaction space for the electrons is small compared to the other circuits. The electron–light interaction in the metal microslit is discussed in detail in Sect. 5.

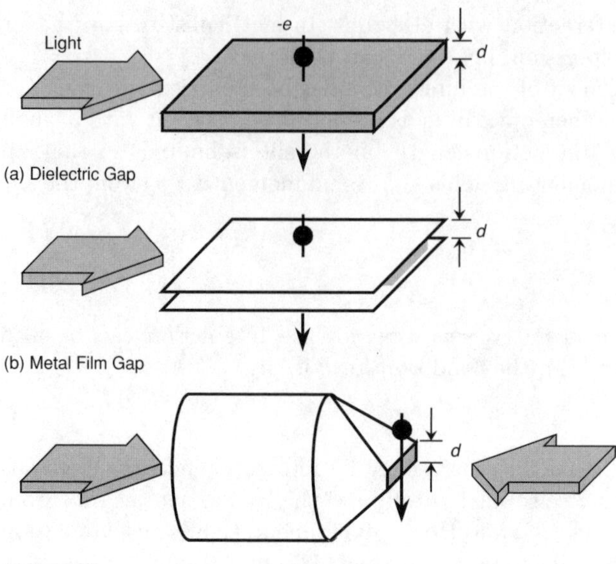

Fig. 5a–c. Schematic drawing of three different gap circuits. The gap width d is smaller than the wavelength of the incident laser

4.2 Transition Rates of Electrons

The transition rates of electrons in a metal film gap and a thin dielectric film were theoretically estimated in [21]. The calculation model is shown in Fig. 6. The electron beam passes through the gap and travels in the $+x$ direction, and the laser wave propagates in the $+y$ direction. To simplify the calculation, the following assumptions were made: (i) the initial velocities of the electrons are identical; (ii) the propagation modes of the light wave in the metal film gap and the thin dielectric film are TEM_{00} and TM_{01}, respectively; (iii) the incident light wave is polarized in the x direction; (iv) the gap materials have no rf loss; and (v) the electrons interact only with the light. The transition rate for an electron in the metal film gap was calculated in accordance with the analyses by Marcuse [22]. The calculated rate w_m is

$$w_m = \frac{2qc\beta^2}{\varepsilon_0 \hbar^2 \omega^4} i P_i \sin^2\left(\frac{\omega d}{2v}\right) , \tag{14}$$

where i is the electron current density, ε_0 is the dielectric constant of free space, and P_i is the power density of the incident light.

In (14), the value of w_m represents the rate at which an electron absorbs one photon with energy $\hbar\omega$. The rate to emit the photon is almost the same as that in (14). Similarly, it is possible to calculate a transition rate w_d for the dielectric film by quantizing the laser field of the fundamental TM_{01}

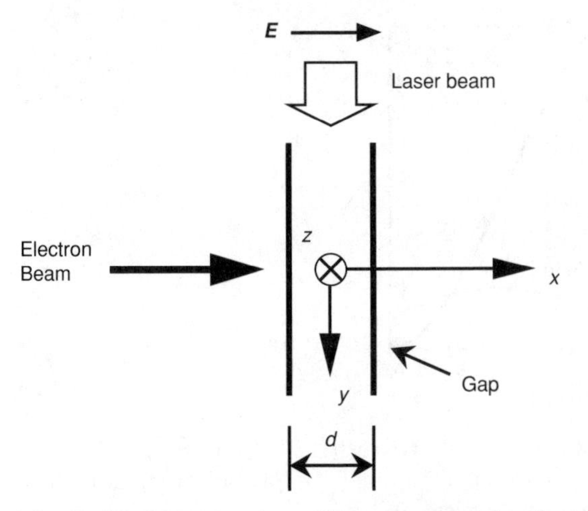

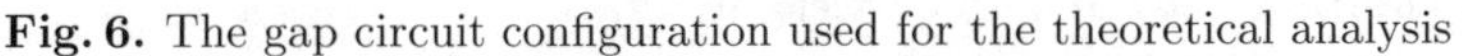
Fig. 6. The gap circuit configuration used for the theoretical analysis

propagation mode. The expression for w_d is

$$w_d = \frac{qc}{\epsilon_0 \hbar^2 \omega^4} i P_i \left| \frac{k_0 k_y d^2}{2{n_i}^2 \cos(k_{ix} d/2)} \right|^2$$

$$\times \frac{\left| 4 \dfrac{k_{ix} d \sin(k_{ix} d/2) \cos(\omega d/2v) - (\omega d/2v) \cos(k_{ix} d/2) \sin(\omega d/2v)}{(k_{ix} d)^2 - (\omega d/v)^2} \right|^2}{\left\{ \dfrac{k_0 d}{2 n_i \cos(k_{ix} d/2)} \right\}^2 + \left\{ \dfrac{({k_y}^2 - {k_{ix}}^2)}{(n_i k_{ix})^2} - \dfrac{{n_i}^2 ({k_y}^2 - {k_{ex}}^2)}{(n_e k_{ex})^2} \right\}}$$

$$\times \frac{1}{\dfrac{k_{ix} d}{2{n_i}^2} \tan(k_{ix} d/2)} , \tag{15}$$

where

$${k_{ix}}^2 + {k_y}^2 = (n_i k_0)^2 ,$$

$${k_{ex}}^2 + {k_y}^2 = (n_e k_0)^2 ,$$

$$|k_{ex}| = \left(\frac{n_e}{n_i} \right)^2 k_{ie} \tan k_{ix} d/2 ,$$

n_i and n_e are the refractive indices inside and outside the dielectric film, respectively, k_y is the wave number in free space in the y direction, and k_{ix} and k_{ex} are the wave numbers inside and outside the film in the x direction. From (14) and (15), the transition rates can be estimated as a function of the gap width. Figure 7 shows the calculated results used to find the optimum gap width. The parameters used in the calculation are $\beta = 0.5$, $\lambda = 780$ nm,

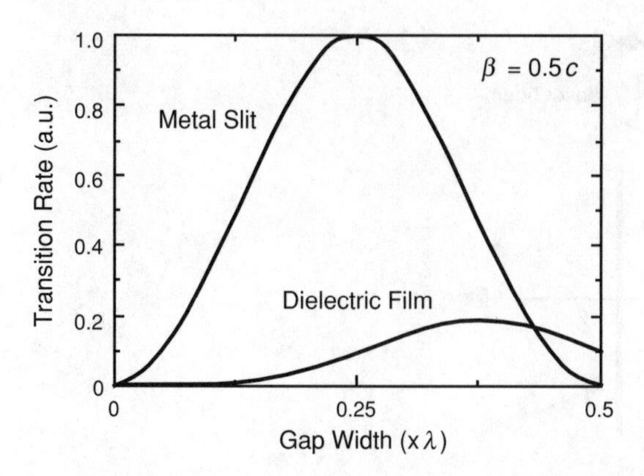

Fig. 7. Calculated transition rates of an electron with a velocity of $\beta = 0.5$ as a function of the gap width of the metal film gap and the SiO_2 dielectric film. The gap width is normalized to the laser wavelength

$n_i = 1.45$ (SiO_2), and $n_e = 1$. As seen from (14), w_m varies sinusoidally and has the peak at the gap width satisfying the equation

$$d = \beta\lambda(m + \frac{1}{2}) \, , \tag{16}$$

where m is an integer. Substituting $\beta = 0.5$ and $m = 0$ for the first peak, an optimum gap width of $\lambda/4$ is obtained. The variation of w_m in Fig. 7 is different from the klystron theory in which the maximum value of w_m is at $d \sim 0$ [5]. The difference between these comes from the different treatments of photon density in the gap. The klystron theory assumes that the total number of photons stored in the gap is constant, but in our treatment, it is assumed that the photon density in the gap is constant and is determined by the incident laser power.

From Fig. 7, it can be seen that for the SiO_2 film, the first peak value of w_d is 0.18 times that of w_m. The optimum film thickness of 0.38 is also longer than the optimum width of the metal film gap. In the dielectric film, the laser field is distributed outside the film as an evanescent wave so that the number of photons inside the gap is smaller than that for the metal gap. The longer gap width increases the number of photons inside the dielectric film.

Figure 8 shows the calculated transition rates as a function of the light intensity for the metal film gap and the SiO_2 film. These transition rates represent the probability per unit time of one electron absorbing a photon. In the calculation, the optimum gap widths of 0.25λ and 0.38λ were used for the metal film gap and the SiO_2 film, respectively. The other calculation parameters were described previously. From Fig. 8, it can be seen that the transition rates are 1.1×10^{-2} and 2×10^{-3}/sec for a power intensity of 10^6 W/cm^2 in the metal film gap and the SiO_2 film, respectively. The power

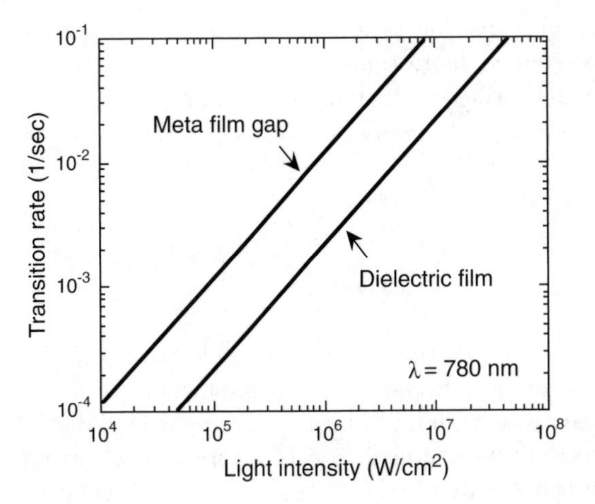

Fig. 8. Calculated transition rates as a function of the power intensity of the incident light for the metal film gap and the SiO_2 dielectric film at a wavelength of 780 nm

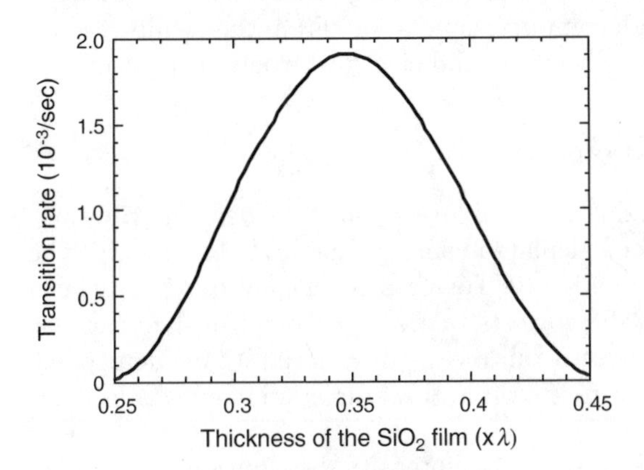

Fig. 9. Transition rate in the SiO_2 dielectric film versus the film thickness at $\lambda =$ 488 nm

intensity of the laser corresponds to an output power of 790 mW focused onto a 10 μm diameter area.

The transition rate just in front of the metal slit is intended to be the same as that of the metal film gap because the field distribution at the surface of the metal slit is similar to that in the film gap. The detailed results for the metal slit are derived via classical theoretical analyses in the next section.

Figure 9 shows the evaluated electron transition rates using (15) for the experimental conditions described by Schwarz and Hora, this is, an Ar laser with a power of 10 W and a beam diameter of 10 μm at $\lambda = 488$ nm, and $v \sim 0.4c$. The optimum film thickness is about 170 nm, which agrees with the one used in the experiment, and is also consistent with the theoretical prediction given in (13). Since the electron transition rate is about 2×10^{-3}/sec, they probably obtained signal electrons at more than 1,000 particles/sec for the electron beam used. Though it is still not clear whether detection of signal electrons with a nonluminescent screen is possible or not, the number of signal electrons may be sufficient to detect the interaction when modern electron detection systems with high sensitivity are used.

5 Theoretical Analyses of a Microslit

Metal microslits can generate near-fields that modulate an electron beam at optical frequencies. The circuit configuration of microslits is suitable for investigating energy exchange of electrons with light, because the near-field distribution on the microslit is precisely determined. In this section, the electron-energy modulation in the microslits is analyzed theoretically in classical terms [23]. Firstly, optical near-field distributions on the slit are determined using the method of moments. The validity of the theory is confirmed by comparing with measured field distributions. The energy changes of electrons are evaluated numerically through computer simulation. From the results, the relationship among wavelength, slit width, and electron velocity is determined.

5.1 Near-Field Distributions

In accordance with Chou and Adams' analysis [24], near-field distributions on a metal microslit have been calculated using the method of moments. The calculation model is shown in Fig. 10. For ease of calculation, it has been assumed that (i) the metal slit consists of two semi-infinite plane screens with perfect conductance and zero thickness, (ii) a normally incident plane wave is polarized perpendicular to the slit in the x direction in which electrons move above the slit.

In order to verify the theory, near-field intensity distributions on a metal slit was measured using a scaled model of the slit at a microwave frequency of 9 GHz (wavelength $\lambda = 33$ mm). The experimental setup is shown in Fig. 11. The metal slit consists of two aluminum plates with a height of 400 mm, a width of 190 mm, and a thickness of 1 mm. A rectangular horn antenna with an aperture size of 116 mm$\times$157 mm was positioned 1650 mm from the slit. This longer distance assures plane wave incidence. A small antenna probe detects the electric field in the x direction, $|E_x|$, which is the dominant field for the interaction with electrons. The antenna probe, with a length of 1.6 mm, terminated a thin coaxial cable with a diameter of 0.8 mm connected to a spectrum analyzer.

Figure 12 compares (a) the calculated and (b) measured field intensity distributions of $|E_x|$ on the slit. The slit width d was 0.5λ. The field intensities were normalized to $|E_x^i|$ measured at $x = y = 0$ without the slit. In Fig. 12, the shapes of the field distributions for both the theory and the experiment are

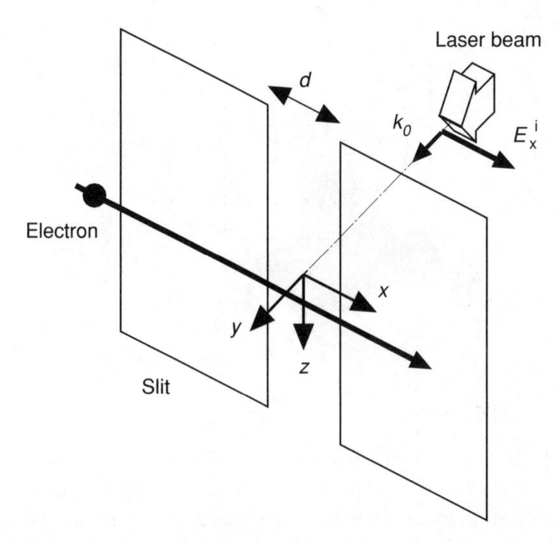

Fig. 10. Calculation model for the near-field distribution in a metal microslit. $|E_x^i|$ denotes the electric field of the incident wave

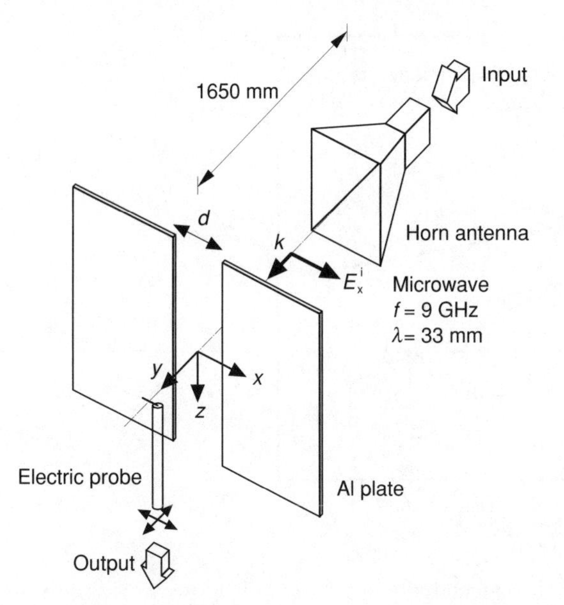

Fig. 11. Experimental setup for measuring the near-field distribution on the metal slit at 9 GHz

quite similar, though the measured near-field in Fig. 12b has small ripples. These would result from interference of waves scattered from the coaxial probe and the slit.

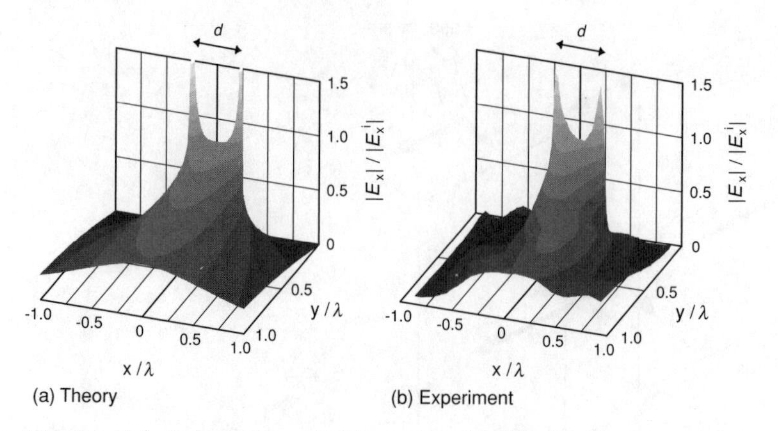

Fig. 12. (**a**) Calculated and (**b**) measured field distributions of the electric field component E_x in the x direction on a slit of width $d = \lambda/4$. E_x is normalized to the incident field intensity E_x^i, and the positions x and y are also normalized to the wavelength

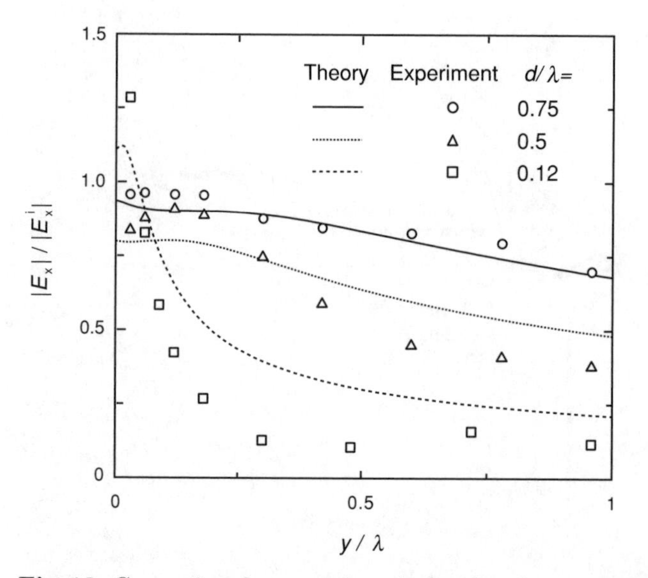

Fig. 13. Comparison between the calculated and measured field intensities for slit widths 0.75λ, 0.5λ and 0.12λ, as a function of the distance from the slit surface y at $x = 0$

Figure 13 shows the calculated and measured field intensities for slits of different width as a function of the distance from the slit surface y at the center of the slit, i.e., $x = 0$. In the results for $d = 0.75\lambda$, the theory agrees well with the measurement. When d decreases from 0.75λ to 0.12λ, deviations of

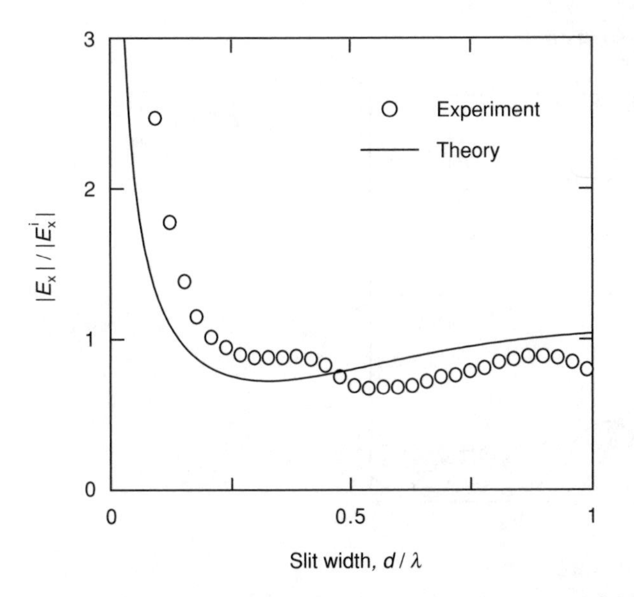

Fig. 14. Comparison between calculated and measured field intensities as a function of normalized slit width d/λ at $x = y = 0$

the measured intensities from the theoretical ones increase mainly due to the probe having a finite size of about 0.05λ. In Fig. 13, the smaller the slit width, the steeper the field decay. The near-fields are localized at a distance within $y \sim d$ from the surface. The near-field distributions on the slit are similar to those on small apertures used as optical near-field probes in scanning near-field optical microscopes [25].

The variation with slit width of the field intensities at $x = y = 0$ has been measured and plotted in Fig. 14. When the slit width decreases from λ to zero, the field intensity is almost the same as that of the incident wave at widths up to about 0.2λ, and then quickly increases. The small ripple in the measured curve might be caused by scattered waves from the coaxial probe as mentioned earlier. The theory predicts the measured variation of field intensities well. Those results indicate that the theory is valid, allowing for experimental errors.

5.2 Wave Number Spectrum

As mentioned in Sect. 3, an optical near-field contains a number of wave components with different wave numbers. We now examine the spatial plane wave spectrum, i.e., the wave number (k) spectrum for the near-field on the slit. Using (6), the near-fields in the spatial domain as shown in Fig. 12 have been transformed to those in the k domain. The transformed spectra for the near-fields E_x at $y = 0.01\lambda$ are shown in Fig. 15. The abscissa is

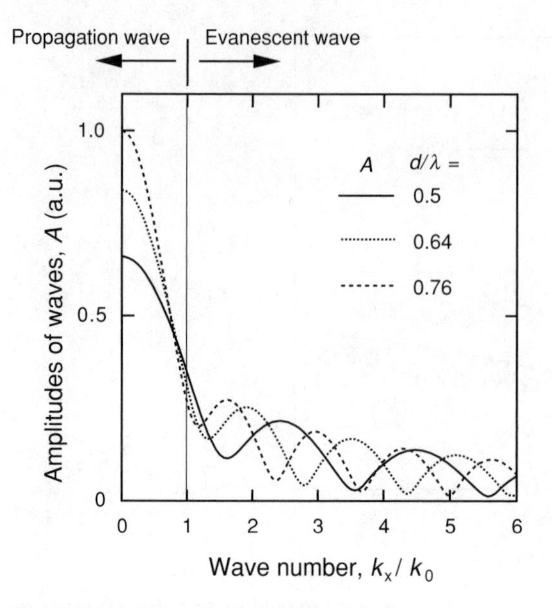

Fig. 15. Calculated amplitudes of the wave components in the near-fields for the slits with widths of 0.5λ, 0.64λ and 0.76λ in the x-direction. The amplitudes A are normalized to the maximum value for the slit with $d/\lambda = 0.76$ at $k_x = 0$

the normalized wave number k_x/k_0 in the x direction. The amplitudes of the wave components, A, are normalized to the maximum value in the curve for $d/\lambda = 0.76$ at $k_x = 0$. The wave components with $k_x > k_0$ in Fig. 15 are evanescent waves that cannot propagate in free space because the wave number $k_y = (k_0{}^2 - k_x{}^2)^{1/2}$ in the y direction is imaginary. As seen from Fig. 15, the k spectra spread over large wave numbers above $6k_0$ and have several peaks at different wave numbers in the evanescent wave region. The wave number k_m where A has a peak shifts toward larger values as d decreases from 0.76λ to 0.5λ. For the first peaks of A in the three spectra, the wave numbers k_m/k_0 are 1.67, 1.92, and 2.38 for slits with $d/\lambda = 0.76$, 0.64, and 0.5, respectively. From the calculation results, it has been found that

$$k_m \cong \frac{2\pi}{d}\left(m + \frac{1}{4}\right) , \tag{17}$$

where m is an integer. This relationship is utilized to design a microslit circuit with the optimum width.

5.3 Numerical Simulations

Using the theoretical near-field distributions, energy changes of electrons passing close to the slit surface were estimated through computer simulation. Referring to Fig. 16, electrons with velocity v move in the x direction at

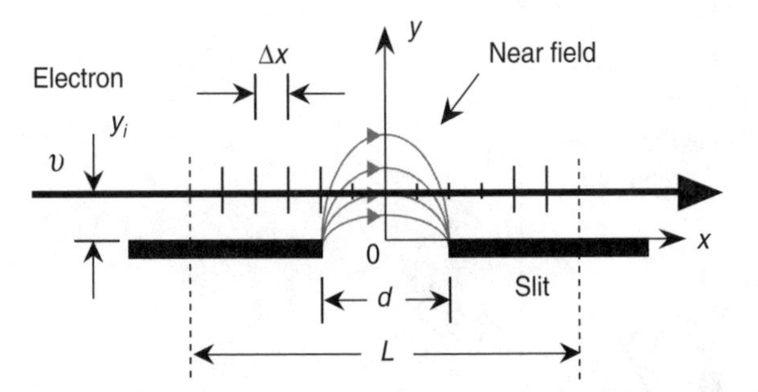

Fig. 16. Calculation model of electron-energy changes with a near-field on a slit of a width d. An electron, with an initial velocity v and its position y_i, is accelerated or decelerated by Lorentz force in the near-field region. The integration length L of ten times the slit width was chosen to fully cover the near-field region on the slit

distance from the slit surface y_i. All field components in the near-field, i.e., electric fields, $\boldsymbol{E}_x$, $\boldsymbol{E}_y$ and a magnetic field, $\boldsymbol{H}_z$, were taken into account for calculation. The total energy variations of the electrons were determined by integrating small energy changes due to the Lorentz force over a small distance Δx along the electron trajectory. A total length L of ten times of the slit width was chosen to entirely cover the near-field region on the slit. In the calculation, a CO_2 laser with power density 10^8 W/cm^2 at $\lambda = 10.6$ µm was assumed as the incident wave. This power density corresponds to a 10 kW output power focused onto a 100 µm diameter area.

Figure 17 shows the calculated energy changes ΔW of electrons passing through the same slits shown in Fig. 15 with various velocities β between 0.8 and 0.2 at $y_i = 0.01\lambda$. The thick lines indicate ΔW and the thin lines are for the k spectra, and are the same as those shown in Fig. 15. ΔW is normalized to the maximum value of the curve for $d = 0.76\lambda$ at $\beta = 0.6$. The two abscissas for k_x and β are related by $\beta = k_0/k_x$, i.e., (10).

As seen from the results in Fig. 17, ΔW is proportional to the amplitude A of the evanescent wave in the near-field on the slit. Those results indicate that electrons interact with a single evanescent wave among a number of wave components contained in the near filed. The computer simulation results thus support the theoretical predictions described in Sect. 3. In Fig. 17, when β decreases, ΔW changes periodically in accordance with the variation of A, and the peaks slowly decrease. In the curve for the slit with $d/\lambda = 0.64$, the peak values of ΔW are 0.93 at $\beta = 0.52$ and 0.61 at $\beta = 0.28$. Those electron velocities correspond to an initial energy of 88 keV and 21 keV, respectively. This fact indicates that the variation of the peaks in the ΔW curves is small compared to that of the initial electron energy.

Using (10) and (17), the slit widths d_m for the peaks of ΔW are given by

$$d_m \cong \beta\lambda(m + \frac{1}{4}) \,. \tag{18}$$

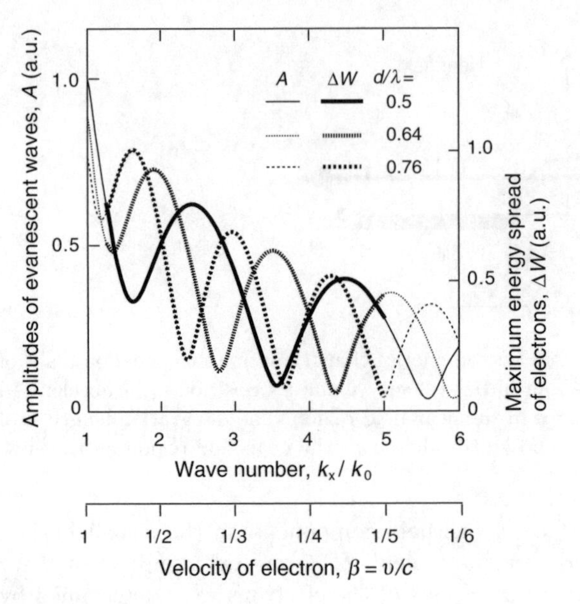

Fig. 17. Calculated energy changes ΔW of electrons passing through the near-fields with the k spectra shown in Fig. 15 with various velocities β at $y_i = 0.01\lambda$. The electron velocities β were chosen to satisfy the relation $\beta = k_0/k_x$ for each of the k spectra. The amplitudes A are normalized to the maximum value in the curve for the slit with $d/\lambda = 0.5$ at $k_x/k_0 = 1$, and ΔW is also normalized to the maximum value of 34 eV for the slit with $d/\lambda = 0.76$ at $\beta = 0.6$

Equation (18) gives the optimum slit width in the metal microslit. Comparing with (16), it is seen that d_m in the microslit is narrower than in the metal film gap by $\beta\lambda/4$. In the metal film gap, electrons are modulated with a uniform field at the film gap. Therefore, the difference between the optimum widths would arise from the difference of the field distributions in the two circuits.

Figure 18 shows the calculated ΔW for various electron velocities as a function of y_i. Slits having the optimum widths $d_m/\lambda = (0.38, 0.5, 0.62)$ for $\beta = (0.3, 0.4, 0.5)$, respectively, were used for calculation. In Fig. 18, when y_i increases from zero to 0.5λ, ΔW falls off exponentially to near zero. Since ΔW is proportional to the field intensity, these curves represent spatial field distributions of the evanescent waves interacting with the electrons. From the evanescent wave theory, the decay constant α of the evanescent wave [26] is given by

$$\alpha = k_0 \sqrt{\left(\frac{k_x}{k_0}\right)^2 - 1}\,. \tag{19}$$

From (10) and (19), the decay constants are estimated to be $\alpha_0 = k_0 \times (3.2, 2.3, 1.7)$ for $\beta = (0.3, 0.4, 0.5)$, respectively, which agree with those estimated from the exponential curves shown in Fig. 18.

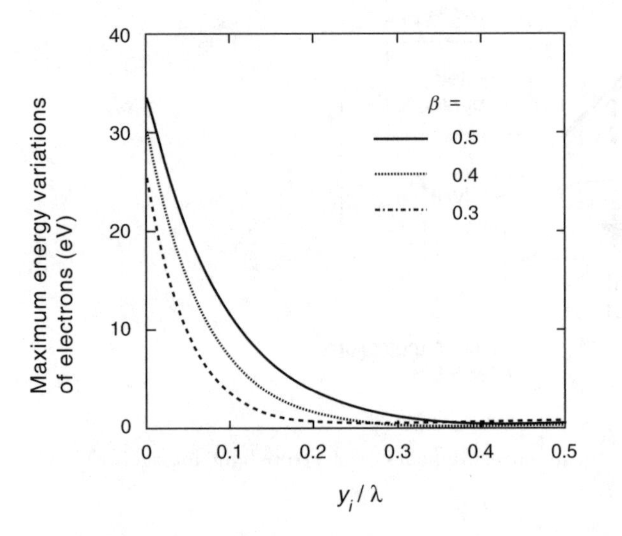

Fig. 18. Calculated maximum electron-energy spreads as a function of the distance y_i/λ from the slit surface for electron velocities $\beta = 0.3$, 0.4, and 0.5

The effective interaction space of the slit can be defined as $y_e = 1/\alpha$, because the field intensity of the evanescent wave falls off by e^{-1}. Using (10) and (19), y_e is

$$y_e = \frac{\lambda}{2\pi}\frac{\beta}{\sqrt{1-\beta^2}} \,. \tag{20}$$

This equation indicates that the interaction space in the slit circuit is highly restricted, particularly for a lower-energy electron beam. The electrons thus must pass very close to the slit surface in order to obtain significant energy exchange with the laser beam. From the ΔW curve for $\beta = 0.5$ in Fig. 18, it is seen that the interaction space for $\lambda = 10.6$ μm is about 3 μm, where measurable electron-energy changes of greater than 1 eV are obtained.

6 Experiment

This section gives experimental verification to the theory. Experiments have been performed in the infrared region, where a wider space on a metal microslit is available for interaction. Electron-energy changes of more than ±5 eV with a 10 kW CO_2 laser pulse at wavelength 10.6 μm has been successfully observed for an electron beam with energy less than 80 keV [27]. The experimental results have been compared with theoretical predictions.

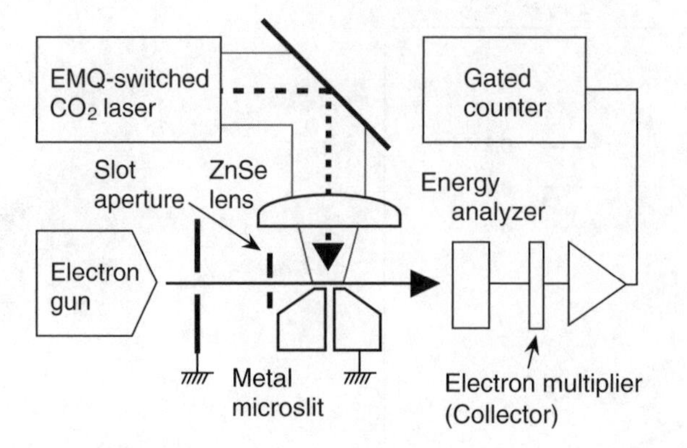

Fig. 19. Experimental setup for measurements of electron–light interaction in the infrared

6.1 Experimental Setup

The experimental setup is shown in Fig. 19. An electromechanical Q-switched (EMQ) CO_2 laser [28] oscillates in the TEM_{00} mode and generates output pulses with maximum peak power 10 kW, width 140 nsec, and repetition rate 1 kpps at $\lambda = 10.6$ µm. The laser beam was focused on the slit surface down to a diameter of about 200 µm using a ZnSe lens. The slit consisted of two polished copper blocks, and the width was 8.4 µm. The initial energy W_i of the electron beam was adjusted between 40 keV and 90 keV. A slot aperture in front of the slit confines the beam area on the slit to 10 µm in height and 100 µm in width. The electron energy was measured using a retarding field analyzer [29]. This analyzer passes all the higher-energy electrons than the filter bias V_f, which is a variable retarding potential.

The pulsed laser output modulates the energy of the electron beam, so that the electron current through the analyzer varies during the pulse. Electrons passed the energy analyzer were detected by a secondary electron multiplier (collector) connected to a gated counter triggered by the laser pulse.

6.2 Electron Energy Spectrum

Figure 20 shows the measured temporal variations of (a) the laser pulse and (b) the corresponding response from the gated counter for an electron beam with $W_i = 80$ keV. The collector current i was 2 pA. The output response was measured by the counter in box-car averager mode with temporal resolution 20 nsec and integration time 0.5 sec. In this experiment, V_f was set to -5.5 V so that the counts in Fig. 20 represent the number of electrons gaining energy more than $| qV_f |$ from the laser.

As seen from Fig. 20, the shape of the output response is considerably different from that of the laser pulse. The response time of 380 nsec in Fig. 20(b)

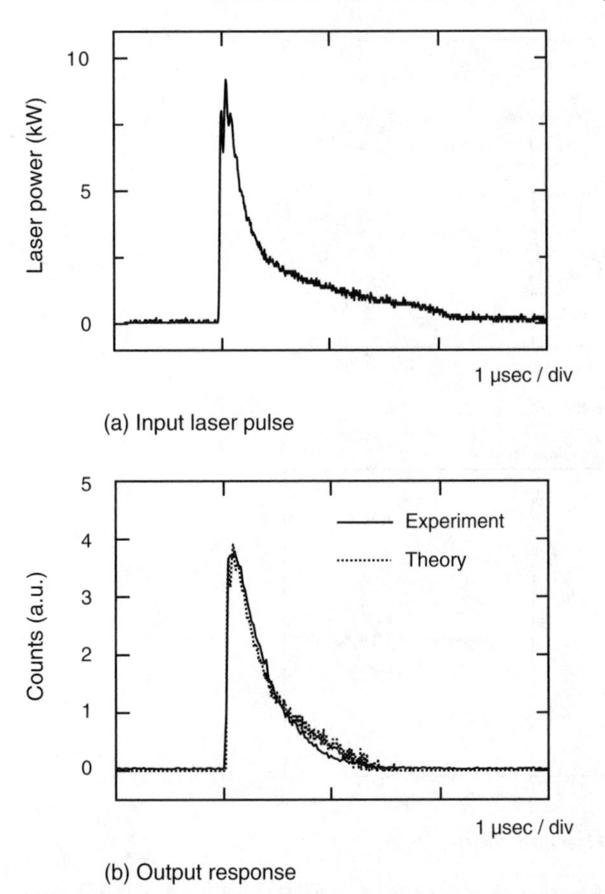

Fig. 20. Measured temporal variation of (**a**) input laser pulse and (**b**) output response from a gated counter for electrons with $W_i = 80$ keV at $\lambda = 10.6$ μm

is much longer than the laser pulse width of 140 nsec. This output response results from the fact that the number of signal electrons is proportional to the laser field, not to the power. In Fig. 20(b), the dotted curve is the theoretical response calculated in a computer simulation described in Sect. 5. In the calculation, the spatial field distribution on the slit has been taken into account. The calculated counts were normalized to the measured peak value. It is seen that the theoretical curve agrees well with the measurement.

Figure 21 shows (a) the measured energy spectra of electrons **A** with and **B** without laser illumination, while (b) shows the difference between the two spectra **A**–**B**. The peak power of the laser was 10 kW and $W_i = 80$ keV. The ordinates are the output counts from the counter with a gate width of 1.5 μsec and an integration time of 10 sec. In Fig. 21(a), the measured spectrum **B** without laser illumination shows that our energy analyzer has

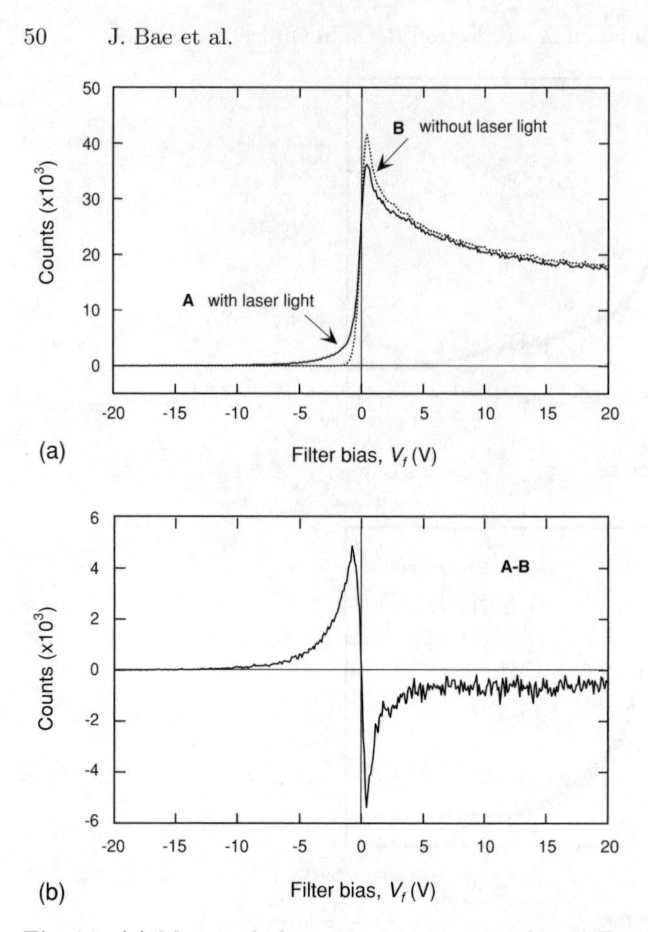

Fig. 21. (**a**) Measured electron-energy spectra **A** and **B**, with and without laser illumination, and (**b**) the difference between the two spectra **A**−**B** for an electron beam with initial energy 80 keV

resolution better than 0.8 eV for an 80 keV electron beam. The output count decreases gradually as V_f increases from +1 eV, due to the dispersion of the energy analyzer. When the laser beam irradiates the electrons, spectrum **B** becomes spectrum **A**, with a wider energy spread. Spectrum **A** still contains a number of electrons that have not interacted with light. In order to remove these electrons and the dispersion effect from the measured spectrum **A**, the output counts in **B** were subtracted from those in **A**. Figure 21b thus indicates the energy spectrum only for electrons that interacted with light. From Fig. 21b, it is seen that the 10 kW laser beam can give an energy spread of more than ±5 eV to the electrons. The experimental results clearly show that using the metal slit, the energy of the electrons can be modulated with a laser at $\lambda = 10.6$ μm.

Since the energy analyzer passes all higher-energy electrons, it is expected that for large bias voltages the output counts with laser illumination should be same as without. However as shown in Fig. 21b, the count with laser illumination is slightly less than without, even at $V_f > +10$ V. This might be due to deflection of the electron beam by the laser illumination. Consequently a part of the electron beam has been clipped by the aperture before the collector.

In Fig. 21, about 70 000 electrons have passed on the slit, and about 13 000 electrons among them have interacted with the laser beam. Since the height of the electron beam on the slit is 10 μm, this ratio of the signal electrons to the total ones implies that the interaction space of the slit is about 2 μm which agrees with the theoretical prediction as described in Sect. 5.3.

6.3 Modulation with Laser Field

In order to accelerate or decelerate electrons with a laser, the laser field must be polarized in the direction of motion of the electrons. Figure 22 shows the measured electron-energy spread as a function of the angle θ between the direction of the electron velocity and the laser polarization. The inset defines θ. The electron-energy spreads ΔW are normalized to the maximum value at $\theta = 0$. The experimental parameters used are the same as the ones described in the previous section. The solid curve is the theoretical variation of ΔW given by $\Delta W \propto E_i \cos\theta$, where E_i is the field intensity of the laser beam.

The variation of the maximum energy spread of electrons with the incident laser power has been measured and is plotted in Fig. 23. In the experiment,

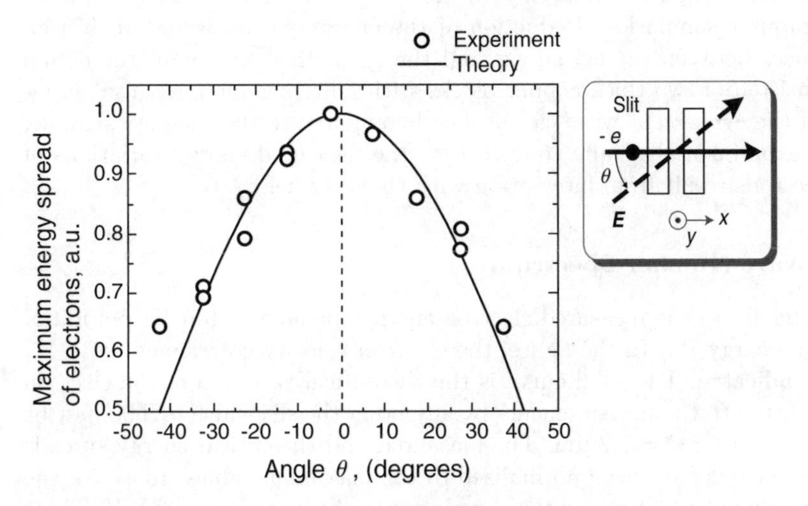

Fig. 22. Measured electron-energy spread versus the polarization angle θ of the incident laser beam. The solid curve is the theoretically predicted variation with θ. In the inset, $\boldsymbol{E}$ is the electric vector of the laser beam

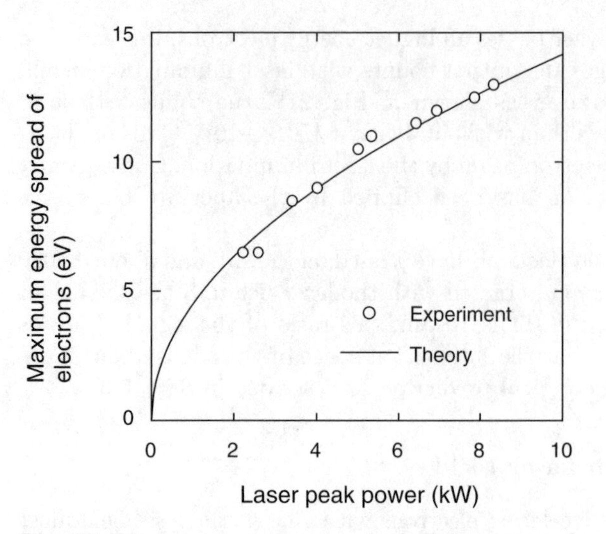

Fig. 23. Maximum energy spread of electrons as a function of laser peak power. The solid curve is a theoretical fit to the measurements

the energy spreads were measured for electrons with $W_i = 80$ keV and a current of 0.5 nA at $V_f < -3$ eV. The solid curve indicates the theoretical variation of the electron-energy, which is proportional to the field intensity of the incident wave, i.e., the square root of the laser power, as mentioned for the results shown in Fig. 20. The theory agrees well with the measurements. The measured electron-energy spread is 13 eV at a laser power of 8.3 kW, which can be compared with the theoretical value of 22 eV predicted via computer simulation. Reduction of the energy spread would arise from differences between the actual slit and the theoretical slit. Since the actual slit consisted of two thick copper blocks with finite conductance, the amplitude of the evanescent wave can be small compared to the theoretical value. Those experimental results confirm that the measured energy variations of the electrons result from interaction with the laser field.

6.4 Wave Number Spectrum

Figure 24 shows the measured electron-energy spread as a function of initial electron energy W_i. In the figure, the electron velocity corresponding to W_i is also indicated. The solid curve is the theoretical variation of the electron energy fitted to the measurements by adjusting the slit width d. The best fit was obtained for $d = 7.2$ μm. The measured and theoretical energy spreads of the electrons have been normalized to the maximum values, 15 eV for the measurements and 34 eV for the theory, respectively, at $W_i = 90$ keV. These experimental results confirm the evanescent wave theory for a metal microslit allowing for experimental errors. In Fig. 24, the electron-energy spread is

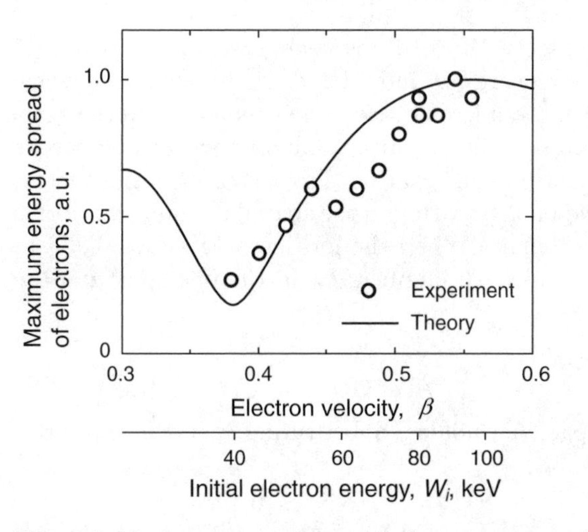

Fig. 24. Maximum energy spread of electrons as a function of initial electron velocity. The solid curve indicates theoretical variations in electron energy in a slit of width 7.2 μm

4 eV at $W_i = 40$ keV. The small modulation at low W_i can be increased by adjusting the slit width. The results indicate that a metal microslit can be used to modulate a nonrelativistic electron beam at optical frequencies. The results shown in Fig. 24 also imply that the k spectrum in an optical near-field on a small object might be measured using an electron beam.

7 Multiple-Gap Circuit

Efficient modulation of an electron beam with a laser can be achieved using an array of metal slits instead of a single microslit. In this section, we describe the interaction between electrons and electromagnetic waves in an interaction circuit with a periodic structure, this is, a metallic diffraction grating. Experiments performed in the far infrared have also verified the evanescent wave theory.

7.1 Inverse Smith–Purcell Effect

In Sect. 6, we described energy variations of electrons induced by an infrared laser in a metal microslit. The maximum energy change with a 10 kW laser beam was about 15 eV. This modulation degree could be raised by using an array of microslits, this is, a diffraction grating. Laser acceleration of electrons using a metal grating is called the inverse Smith–Purcell effect [13].

Figure 25 shows the configuration for the inverse Smith–Purcell effect. When a metal grating is illuminated by a laser with wavelength λ, a near-field is induced on the surface of the grating. Electrons passing through the grating with velocity v interact with an evanescent wave in the near-field region and consequently their energy is modulated at the laser frequency. Similarly to the interaction in the microslit, the evanescent wave in the grating must satisfy the interaction condition, (10), to interact with the electrons because of energy and momentum conservation. From Fourier optics theory, it is known that the near-field on the grating in the spatial domain transforms to a line spectrum in the k domain. When the incident angle θ of the laser beam is taken into account, the wave numbers k_n in the line spectrum are given by

$$k_n = k_0 \cos\theta + \frac{2n\pi}{D} , \tag{21}$$

where n is the number of space harmonics. Substituting k_n for k_e into (10), we obtain

$$\lambda = \frac{D}{n}\left(\frac{1}{\beta} - \cos\theta\right) . \tag{22}$$

This equation is exactly the same as (1) in the Smith–Purcell effect in which electrons emit light. This is the reason why the effect shown in Fig. 25 is called the inverse Smith–Purcell (ISP) effect.

The relation $k_e = \omega/v$ in (10) can be written $v = v_p$, where $v_p(= \omega/k_e)$ is the phase velocity of the evanescent wave. The interaction condition in (10) and (22) is thus also called the synchronous condition or the phase-matching condition. The above discussion shows that the basic principle of the interaction is the same for both a metal microslit and a grating. Therefore, the theoretical prediction for the field distributions of evanescent waves described in Sect. 5 has been confirmed by measuring the ISP effect in the far infrared [30].

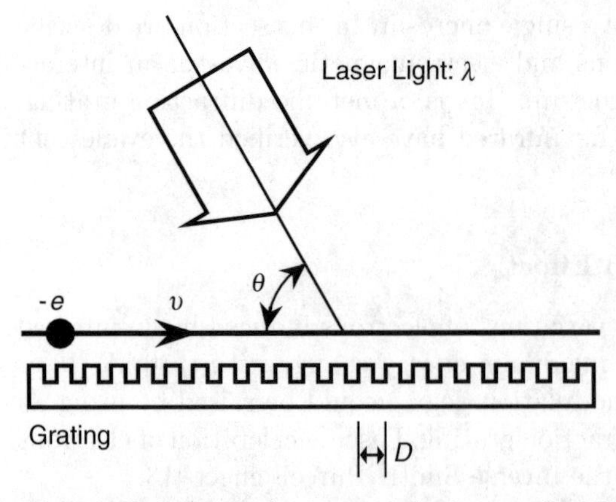

Fig. 25. Schematic drawing of the inverse Smith–Purcell effect. D is the period of the grating, d is the gap width, v is the electron velocity, and θ is the incident angle of the laser light

7.2 Experimental Setup

Experiments have been carried out using a far-infrared laser at 496 μm [31]. Figure 26 shows the experimental setup. A submillimeter wave (SMM) gas laser with a longer wavelength provides a wide enough interaction space on the grating to measure the field distribution of the evanescent waves precisely. In Fig. 26, the metal grating has gaps with a rectangular cross section, the pitch is 246 μm, and the gap width and depth are 40 μm and 104 μm, respectively. The laser was an optically pumped CH_3F laser [32] which oscillated in the fundamental (TEM_{00}) mode and had the pulsed output with peak power between 1 and 80 W. Two lenses focused the laser beam to the grating surface at $\theta = \pi/2$ (refer to Fig. 25). The spot sizes on the surface were calculated to be 1.2 mm for the CH_3F laser on the basis of Gaussian beam theory.

The electron energy analyzer is the same as the one described in Sect. 6.1. The secondary electron multiplier and the gated counter were replaced by a metal plate collector and a boxcar averager because larger numbers of signal electrons were expected in the experiment. A movable slit with a gap of 10 μm was placed at the end of the grating, and was used to specify the position of the electron beam above the grating to an accuracy of ±3 μm.

7.3 Phase Matching Condition

Figure 27 shows the experimental results for the phase matching condition between the electron velocity and the phase velocity of the evanescent waves in the first and second space harmonics, i.e., $n = 1$ and 2 in (22). The laser power was between 8.3 and 12 W. The ordinate is electron energy spread normalized to its maximum value. The abscissa is the initial electron energy W_i. As the interaction length between electrons and evanescent waves on the grating is finite, an effective interaction can occur for electrons with a certain

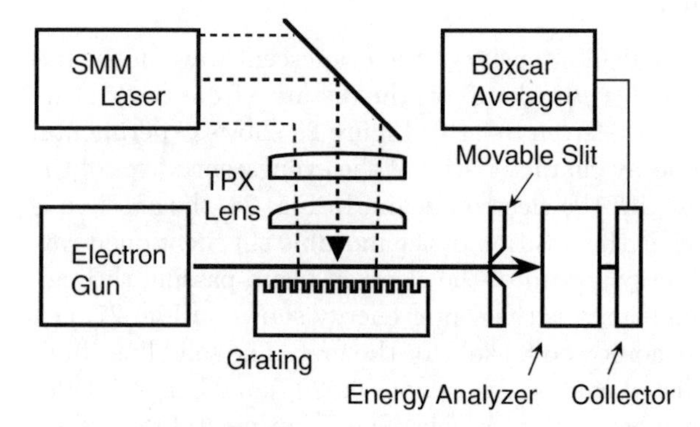

Fig. 26. Experimental setup for measuring the inverse Smith–Purcell effect. The SMM laser is a submillimeter wave laser and TPX is poly 4 methylpenten-1 which is a low-loss material in the submillimeter wave region

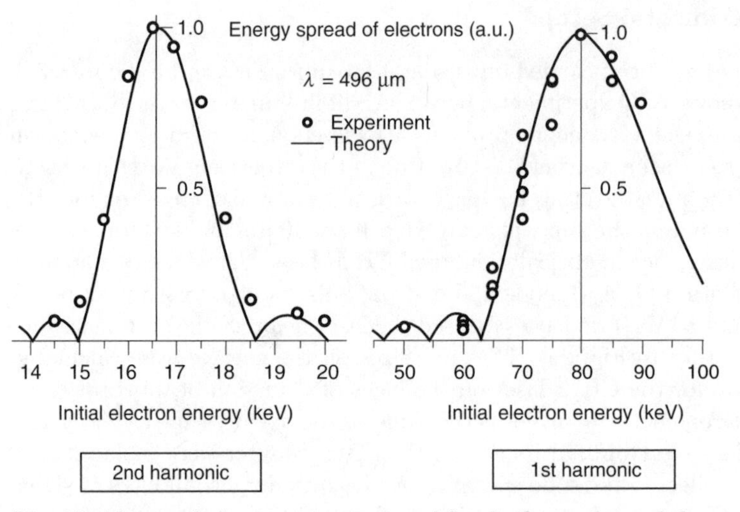

Fig. 27. Measured electron-energy spread for the first order and the second order of space harmonics as a function of the initial electron energy. The points represent the experimental values normalized to the maximum values and the solid lines are the theoretically predicted variation in energy spread for an interaction length of 3 mm

range. The largest energy spreads are produced at $W_i = 16.6$ keV for the second harmonic and 80 keV for the first harmonic, respectively, which can be compared to 16.5 keV and 77.5 keV estimated from (22). By curve fitting, we can deduce that the effective interaction length is 3 mm. Theoretical plots for the length are given by the solid curves in Fig. 27 [33]. These experimental results show that the theoretical considerations for the electron–light interaction in the grating is valid, allowing for experimental errors.

7.4 Field Distributions

As described in Sect. 5.3, the field intensity of the evanescent wave in the microgap is proportional to $\exp(-\alpha y)$, where y is the distance from the grating surface. The decay constant α is given by (19). Figure 28 shows experimental results that show the field decay characteristics of the evanescent waves of the first and second harmonics with the electron beam. In Fig. 28, the abscissa is the electron position, which is the position of the movable slit mentioned earlier. The ordinate is the energy spread of the electron beam passing through the slit. The initial electron energy is the center energy shown in Fig. 27, i.e., 16.6 keV for the second harmonic and 80 keV for the first. The solid lines indicate the theoretically predicted changes, i.e., $\exp(-0.022y)$ and $\exp(-0.049y)$ for the first and second harmonics, respectively. The experimental results are in good agreement with theory. The experimental results are a direct verification of the evanescent wave theory for the inverse Smith–Purcell effect, and thus for the metal microslit interaction circuit.

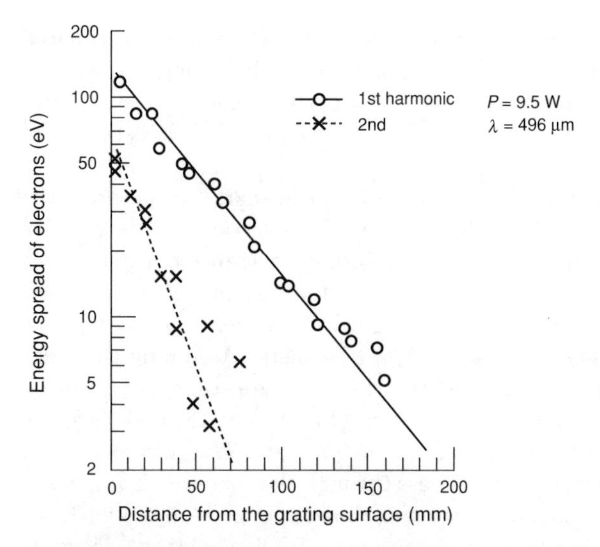

Fig. 28. Measured energy spread of electrons interacting with the first order and second order of space harmonics as a function of distance from the grating surface. The solid and dotted lines are the theoretically predicted variations, and P is the incident laser power

8 Microslit for Visible Light

Figure 29 shows a conceptual drawing of the experimental system for the interaction of electrons with light at shorter wavelengths. A metal microslit is fabricated at the end of an optical fiber so that the laser beam is guided to the slit without requiring precise adjustment. When a laser beam with photon energy greater than 1 eV is used for the interaction, the quantum effects are detectable because the electron-energy analyzer can resolve energy changes in electrons due to photons. Based on the theoretical and experimen-

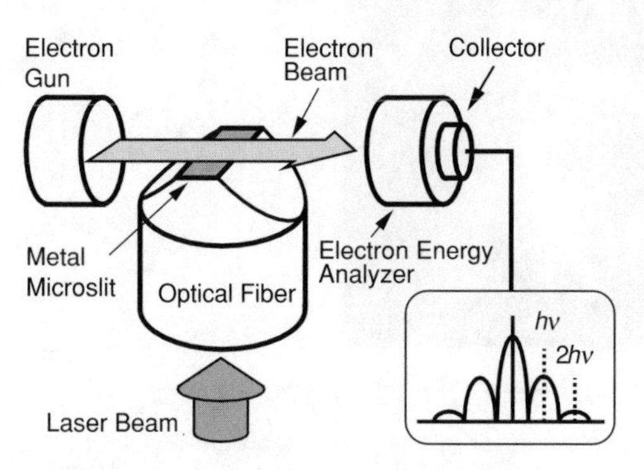

Fig. 29. Conceptual drawing of the experimental setup for measurements of the electron–light interaction in near-infrared and visible light regions

tal results, the number of signal electrons in the interaction was estimated in both classical and quantum treatments, assuming a laser output power of 30 mW at $\lambda = 780$ nm, and an electron beam with velocity $\beta = 0.5$ and density 1 $\mathrm{mA/cm^2}$. It was also assumed that the diameter of the laser beam in the fiber was 6 μm.

Using the classical theory described in Sect. 5, the energy variation of the electrons was estimated to be less than 0.2 eV. This is too small to observe the interaction in practical experiments. In order to transfer a detectable energy of 1 eV to the electrons, a laser power of more than 1 W is required.

In contrast to the conclusion in the classical consideration, a different result is derived from a quantum mechanical treatment. Assuming the transition rates in the metal film gap and the microslit are same, a transition rate of about 1×10^{-3}/sec was obtained using (14). From (18) and (20), the optimum slit width and interaction space are 490 nm and 72 nm, respectively. Then 2.7×10^{7} electrons per second can pass through the space of 72 nm×6 μm on the slit. The number of signal electrons is thus more than 20 000 particles per second. This number is enough to experimentally demonstrate electron–light interaction. Since the interaction behavior between electron and light in a transition region from classical to quantum regime is still vague, the final conclusion for the above discussions should be based on experimental evidence.

In the experiment in the visible light region, the key device is a metal microslit with submicron width. Figure 30 shows a prototype of the microslit fabricated at the center of the core of an optical fiber with 125 μm diameter utilizing a chemical etching technique [34]. In Fig. 30, (a) is a ridge structure and (b) is the same ridge with aluminum coating and a 270 nm gap at the top. The ridge structure is needed to avoid collision of the electrons with

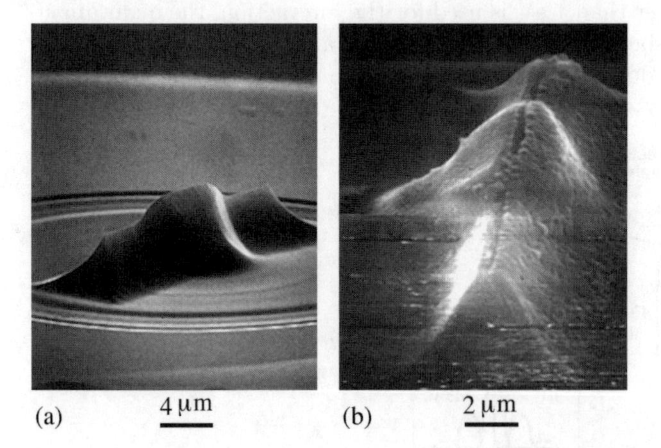

Fig. 30. Metal micro slit fabricated at the center of the end of a single-mode optical fiber: (**a**) ridge structure with a taper angle of 80°, (**b**) metal slit with a width of 270 nm fabricated on the ridge structure

the slit due to the image force. The dimensions of the ridge structure are 5.5 μm in length and 7.2 μm in height. The taper angle is 80°. Both the flatness along the length of ridge and the radius of curvature at the top of the ridge are less than 30 nm. Another probe with a 180 nm-width slit was fabricated using gold instead of aluminum as a coating metal. Those metal microslits can be used for measurements of the interaction in the visible light region. The metal microslits with a narrower width could be fabricated in a way similar to how conventional near-field probes with nanometric apertures have been fabricated [35].

9 Conclusion

Energy modulation of nonrelativistic electrons with optical near-fields has been discussed. A theory based on Fourier optics has shown that the electrons exchange energy with an evanescent wave contained in the near field when the phase velocity of the wave is equal to the velocity of the electrons. This interaction condition derived from the evanescent wave theory is also consistent with conservation of energy and momentum in the interaction. A metal microslit has been adopted to generate optical near-fields with laser illumination. For the interaction in the microslit, the relationship among slit width, electron velocity, and wavelength has been found through theoretical analyses based on computer simulation. Those theoretical predictions have been verified experimentally in the infrared. Electron energy changes of more than ±5 eV with a 10 kW CO_2 laser pulse at a wavelength of 10.6 μm has been observed for an electron beam with an energy of less than 80 keV. From the experimental and theoretical results, it can be concluded that the microslit could be used to investigate physical processes involving in electron–light interaction including its quantum effects in the visible light region. The research results will contribute to developing new types of optical near-field microscopes which measure wave-number distributions in the near-field of a nano-object by using an electron beam.

References

1. S. Okamura: *History of Electron Tubes.* (IOS press, Inc., Washington DC 1994)
2. C.A. Spindt, I. Brodie, L. Humphrey, E.R. Westerberg: J. Appl. Phys. **47**, 5248 (1976)
3. J. Mohr, C. Burbaum, P. Bley, W. Menz, U. Wallrabe: *Micro System Technologies.* ed. by H. Reichl (Springer, Berlin Heidelberg New York 1990)
4. K. Yokoo, M. Arai, M. Mori, J. Bae, S. Ono: J. Vac. Sci. Technol. B **13**, 491 (1995)
5. R.G.E. Hutter: *Beam and Wave Electronics in Microwave Tubes.* (D. Van Nostrand, Toronto 1960)
6. C.A. Brau: 'Free-Electron Lasers'. In: *Advances in Electronics and Electron Phys., Suppl. 22.* (Academic Press, New York 1990)

7. J. Lecante, Y. Ballu, D.M. Newns: Phys. Rev. Lett. **38**, 36 (1977)
8. H. Cohen, T. Maniv, R. Tenne, Y. Rosenfeld Hacohen, O. Stephan, C. Colliex: Phys. Rev. Lett. **80**, 782 (1998)
9. I.R. Senitzky: Phys. Rev. **95**, 904 (1954)
10. S.J. Smith, E.M. Purcell: Phys. Rev. Lett. **92**, 1069 (1953)
11. P.M. van den Berg: J. Opt. Soc. America **63**, 1588 (1973)
12. M. Goldstein, J.E. Walsh, M.F. Kimmitt, J. Urata, C.L. Platt: Appl. Phys. Lett. **71**, 452 (1997)
13. K. Mizuno, S. Ono, O. Shimoe: Nature **253**, 184 (1975)
14. Y. Takeda, I. Matsui: Nucl. Instrum. Methods **62**, 306 (1968)
15. R.B. Palmer: Particle Acceleratiors **11**, 81 (1980)
16. J.M. Wachtel: J. Appl. Phys. **50**, 49 (1979)
17. H. Schwarz, H. Hora: Appl. Phys. Lett. **15**, 349 (1969)
18. H. Hora, P.H. Handel: 'New Experiments and Theoretical Development of the Quantum Modulation of Electrons (Schwarz-Hora Effect)'. In: *Advances in Electronics and Electron Physics.*. ed. by P.W. Hawkes (Academic Press, New York 1987) pp. 55–113
19. R.H. Pantell: 'Interaction between Electromagnetic Fields and Electrons'. In: *AIP Conference Proceedings No. 87, Physics of High Energy Particle Accelerators.* ed. by R.A. Crrigan, F.R. Huson (American Institute of Physics, New York 1981) pp. 863–918
20. G.A. Massey: Appl. Opt. **23**, 658 (1984)
21. J. Bae, S. Okuyama, T. Akizuki, K. Mizuno: Nucl. Instrum. Methods **331**, 509 (1993)
22. D. Marcuse: *Engineering Quantum Electrodynamics.* (Academic Press, New York, 1970) pp. 127–142
23. R. Ishikawa, J. Bae, K. Mizuno: J. Appl. Phys. **89**, 4065 (2001)
24. T.Y. Chou, A.T. Adams: IEEE Trans. Electromagn. Compat. **19** 65 (1977)
25. Y. Leviatan: J. Appl. Phys. **60**, 1577 (1986)
26. D.P. Tsai, H.E. Jackson, R.C. Reddick, S.H. Sharp, R.J. Warmack: Appl. Phys. Lett. **56**, 1515 (1990)
27. J. Bae, R. Ishikawa, S. Okuyama, T. Miyajima, T. Akizuki, K. Okamoto, K. Mizuno: Appl. Phys. Lett. **76**, 2292 (2000)
28. J. Bae, T. Nozokido, H. Shirai, H. Kondo, K. Mizuno: IEEE J. Quantum Electron. **30**, 887 (1994)
29. J.F. Graczyk, S.C. Moss: Rev. Sci. Instrum. **40**, 424 (1969)
30. K. Mizuno, J. Bae, T. Nozokido, K. Furuya: Nature **328**, 45 (1987)
31. J. Bae, H. Shirai, T. Nishida, T. Nozokido, K. Furuya, K. Mizuno: Appl. Phys. Lett. **61**, 870 (1992)
32. M.S. Tobin: Proc. IEEE **73**, 61 (1985)
33. J. Bae, K. Furuya, H. Shira, T. Nozokido, K. Mizuno: Jpn. J. Appl. Phys. **27**, 408 (1988)
34. J. Bae, T. Okamoto, T. Fujii, K. Mizuno: Appl. Phys. Lett. **71**, 3581 (1997)
35. M. Ohtsu: J. Lightwave Tech. **13**, 1200 (1995)

Fluorescence Spectroscopy with Surface Plasmon Excitation

T. Neumann, M. Kreiter, and W. Knoll

1 Introduction

In recent years, much effort has been directed towards the development of optical biosensors. While direct sensors are capable of monitoring the presence of an analyte without the use of labelling groups, the class of indirect sensors exploits the signal enhancement caused by bound marker molecules. Surface plasmon spectroscopy (SPS) as a direct detection method [1] is known to lack sensitivity for monitoring of low molecular mass analytes. In order to enhance the sensitivity and to improve the detection limit the technique was combined with fluorescence detection schemes in surface plasmon fluorescence spectroscopy (SPFS), as described recently [2]. Here, we briefly review the theory of plasmon excitation and the experimental realization of SPFS.

2 Theoretical Considerations

2.1 Surface Plasmons at the Interface Between a (Noble) Metal and a Dielectric Medium

The phenomenon of surface plasmons has been known for a long time. The underlying principles and theories are well understood, and a number of publications can be found which discuss their properties in detail [3–5]. Surface plasmons are surface waves which can be excited at the interface between a metal and a dielectric, and the exact excitation conditions strongly depend on the optical properties of the system. We will see that changes in these properties will lead to altered experimental excitation conditions. This measurable response of the system permits the sensitive monitoring of processes near this interface.

Surface plasmons (Plasmon Surface Polaritons, PSP) are collective oscillations of the quasi-free electron gas of a metal propagating along the interface to a dielectric medium. The field intensity of this evanescent surface wave reaches into both media (Fig. 1); however, it decays exponetntially. PSP also dissipate energy in the metal and, hence, are damped along their propagation direction x. Before the various excitation mechanisms for surface plasmons are discussed, the theoretical background of electromagnetic waves and their interaction with metallic surfaces is discussed in the following.

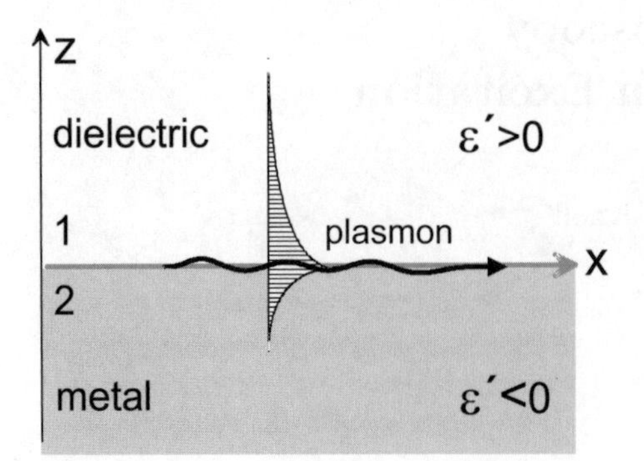

Fig. 1. Schematic representation of a surface plasmon at the interface between a metal and a dielectric. Note that the dielectric constants of the two media need to be of opposite sign in order to allow for plasmon excitation

In order to solve Maxwell's equations for surface plasmons we have to consider an interface between two media. In general, the dielectric constants of both adjacent media can be complex:

$$\tilde{\varepsilon}_1 = \varepsilon_1' + \mathrm{i}\varepsilon_1'' , \qquad \tilde{\varepsilon}_2 = \varepsilon_2' + \mathrm{i}\varepsilon_2'' . \tag{1}$$

If we assume that the magnetic permeability equals 1, we find the following connection between the dielectric constant $\tilde{\varepsilon}$ and the complex refractive index $(n + \mathrm{i}\kappa)$ of the materials. The two descriptions are equivalent and can be converted into each other:

$$(n+\mathrm{i}\kappa)^2 = \varepsilon' + \mathrm{i}\varepsilon'' = \tilde{\varepsilon} , \quad \varepsilon' = n^2 - \kappa^2 , \quad \varepsilon'' = 2n\kappa \tag{2}$$

with n being the refractive index of the material; κ is the absorption coefficient, which describes the damping of the propagating electromagnetic wave due to interactions with the material.

In order to describe the electromagnetic wave in both media the Maxwell equations of the following form should be solved:

$$\left.\begin{array}{l} \boldsymbol{E}_1(\boldsymbol{r},t) = \boldsymbol{E}_1 \cdot \exp(\mathrm{i}(\tilde{k}_{x1}x + \tilde{k}_{z1}z - \omega t)) \\ \boldsymbol{H}_1(\boldsymbol{r},t) = \boldsymbol{H}_1 \cdot \exp(\mathrm{i}(\tilde{k}_{x1}x + \tilde{k}_{z1}z - \omega t)) \end{array}\right\} \text{for } z > 0$$
$$\left.\begin{array}{l} \boldsymbol{E}_2(\boldsymbol{r},t) = \boldsymbol{E}_2 \cdot \exp(\mathrm{i}(\tilde{k}_{x2}x + \tilde{k}_{z2}z - \omega t)) \\ \boldsymbol{H}_2(\boldsymbol{r},t) = \boldsymbol{H}_2 \cdot \exp(\mathrm{i}(\tilde{k}_{x2}x + \tilde{k}_{z2}z - \omega t)) \end{array}\right\} \text{for } z < 0 , \tag{3}$$

where $\tilde{k}_{x1}$, $\tilde{k}_{x2}$ and $\tilde{k}_{z2}$ are the wave vectors for media 1 and 2 in the x and z direction, respectively. Due to the symmetry of the problem a separation of the electromagnetic fields into transverse magnetic (TM, p-polarization) modes with $\boldsymbol{H} = (0, H_y, 0)$ and transverse electric (TE, s-polarization) modes with $\boldsymbol{E} = (0, E_y, 0)$ is possible. All other solutions are superpositions of these two cases.

Since the electric field vector in s-polarization (TE) has no component perpendicular to the interface, no surface polaritons can be excited. Therefore, the excitation of plasmons is not possible by s-, but only with p-polarized light. Since the tangential components of $\boldsymbol{E}$ and $\boldsymbol{H}$ are steady at the interface, it follows that

$$E_{x1} = E_{x2}\ , \quad H_{y1} = H_{y2} \tag{4}$$

and, therefore, a comparison with (3) leads to $k_x = k_{x1} = k_{x2}$. Together with $c_0 = 1/\sqrt{\varepsilon_0 \mu_0}$ and by insertion of (3) into Maxwell's equations one obtains

$$\tilde{k}_{z1} H_{y1} - \frac{\omega}{c_0} \tilde{\varepsilon}_1 E_{x1} = 0\ , \quad \tilde{k}_{z2} H_{y2} - \frac{\omega}{c_0} \tilde{\varepsilon}_2 E_{x2} = 0\ . \tag{5}$$

Finally, insertion of (4) leads to the only nontrivial solution:

$$\frac{\tilde{k}_{z1}}{\tilde{k}_{z2}} = -\frac{\tilde{\varepsilon}_1}{\tilde{\varepsilon}_2}\ . \tag{6}$$

In other words, such an electromagnetic wave can only exist at the interface between two media that have dielectric constants of opposite sign. Therefore, plasmons can be excited at the interface between metal layers (negative ε') and dielectrics. Frequently, silver is used as the metallic layer, but for many practical purposes, gold is preferred due to its high stability against most buffer solutions and environmental influences.

Finally, the dispersion relation of surface plasmons can be derived from the given equations:

$$\tilde{k}_x = k'_x + \mathrm{i} k''_x = \frac{\omega}{c_0} \sqrt{\frac{\tilde{\varepsilon}_1 \tilde{\varepsilon}_2}{\tilde{\varepsilon}_1 + \tilde{\varepsilon}_2}}\ . \tag{7}$$

The amplitude of the electric field decays exponentially into both media in the z-direction as well as in the propagation direction. The propagation length can be calculated by

$$L_x = \frac{1}{2k''_x} \tag{8}$$

and plays an important role for the lateral resolution of methods like surface plasmon microscopy. The length L_x limits the dimensions of the smallest structures that can be detected by this method. For an interface between gold ($\tilde{\varepsilon} = -12 + \mathrm{i} \cdot 1.3$) and a dielectric of $\varepsilon \approx 2.25$, and a laser wavelength $\lambda = 632.8$ nm, the propagation length is about $L_x = 5$ µm.

2.2 Optical Excitation of Surface Plasmons

Although it is known that surface plasmons can be excited by electrons, excitation by incident transverse magnetic (TM) light is the frequently used method of choice, due to the ease in experimental handling. In addition to the limitation to p-polarized light, a number of additional conditions have to be fulfilled in order to excite surface plasmons.

Consider the dispersion relation of free photons travelling through a dielectric medium. Then we find that at each frequency the wave vector for photons is always less than that for surface plasmons propagating along the metal–dielectric interface:

$$k_{\mathrm{ph}} = \frac{\omega}{c}\sqrt{\tilde{\varepsilon}} < \frac{\omega}{c}\sqrt{\frac{\tilde{\varepsilon}_1\tilde{\varepsilon}_2}{\tilde{\varepsilon}_1+\tilde{\varepsilon}_2}} = k_{\mathrm{psp}} \,. \tag{9}$$

A schematic representation of this relation is given in Fig. 2. Although the dispersion relation of the plasmons (P1) approaches the linear relation of free photons (a), there is no intersection of the two curves. The dark grey shaded area represents the frequency/wavevector combinations that are accessible to light travelling in the dielectric medium. It is therefore important to note that surface plasmons are not directly excitable by light, unless the wave vector of the free photons is increased by appropriate experimental techniques, which are described in the following.

Prism Coupling

Although surface plasmons exhibit photonic character, direct coupling of incident light is not possible on a smooth surface, because the momentum of light in a dielectric medium is too low to allow for coupling to the surface mode. One method of increasing the momentum of the photons is to pass them through a medium with higher refractive index than that of the proximal dielectric. This can be done by reflecting the incident light off the base of a high refractive index prism (Fig. 3b). In this so-called Kretschmann prism coupling [6], the surface plasmons are excited by the evanescent field of the reflected light under total internal reflection conditions. A detailed description of prism coupling can be found in the literature [1,3,7,8].

A thin metal film is evaporated directly onto the base of the prism and plasmon excitation at the metal/dielectric interface can be used to sense changes in the adjacent dielectric medium. The maximum thickness of the metal film is limited, because metal films that are too thick damp the incoming beam – excitation of plasmons on the opposite site of the metal film is not possible. On the other hand, if the film is too thin, losses of the excited plasmon mode occur via the field reaching back into the prism. Therefore, an optimum thickness exists for gold in the range $d = 48$ nm under excitation by $\lambda = 623.8$ nm laser light.

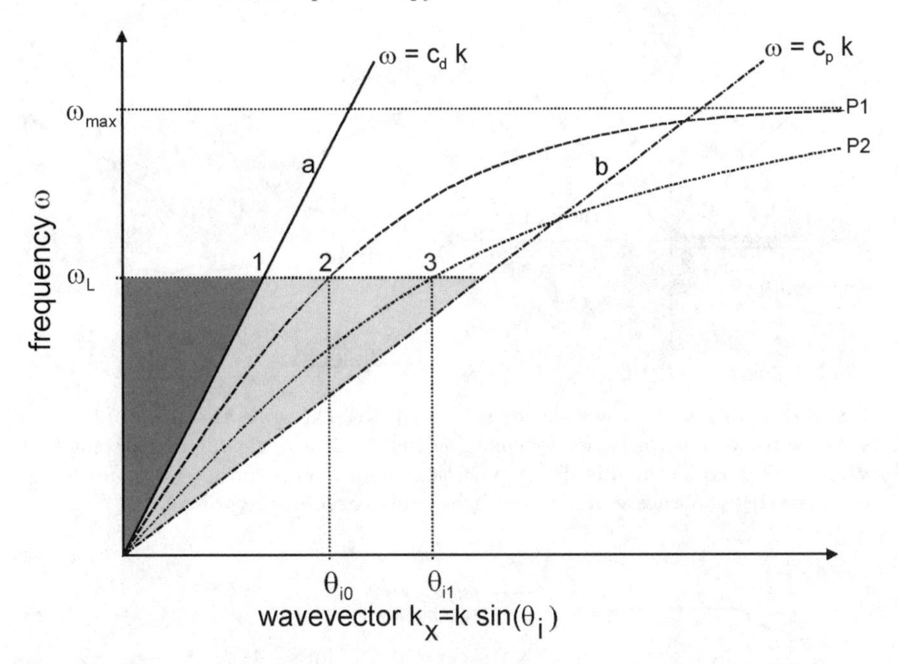

Fig. 2. Dispersion relation of (a) free photons in a dielectric medium, (b) free photons propagating in the coupling prism, compared with the dispersion relation for surface plasmons at the interface between metal and dielectric, before (P1) and after (P2) adsorption of an additional dielectric layer. At a given laser wavelength (ω_1), energy and momentum matching between photons in the prism (b) and plasmons can be achieved (2,3) due to an increase in the wave vector of the free photons (b)

As schematically illustrated in Fig. 2b, the dispersion relation of free photons that pass the high refractive index prism (b) is changed compared to those passing through the dielectric: the momentum of the photons, k_{ph}, at a given frequency ω_L is increased compared to photons outside the prism (a). Due to this increase, all k values in the light grey shaded area are accessible now to the reflected light, and a match with the dispersion relation of the surface plasmons is possible.

The important quantity for the excitation of surface plasmons is the projection $k_{\text{ph},x}$ of the wave vector of the light k_{ph} along the propagation direction x of the surface waves. It is possible to tune the excitation into resonance by changing the incidence angle θ of the incoming laser beam until

$$k_{\text{ph},x} = k_{\text{ph}} \sin(\theta) \tag{10}$$

matches the surface plasmon wave vector k_{psp} and resonance occurs. This situation corresponds to the intersection 2 in Fig. 2, where resonant coupling occurs at the given combination of the laser frequency and incidence angle. Once the system is in resonance, surface plasmons are excited by the incident laser light, resulting in a decrease in reflected intensity. Therefore, a

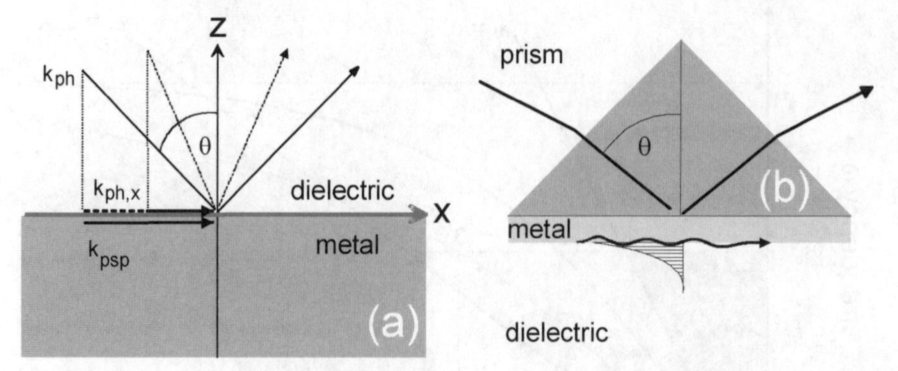

Fig. 3. Excitation of surface plasmons by light. (**a**) The projection $k_{ph,x}$ of the light wave vector k_{ph} can be varied by changing the incident angle θ, but matching with k_{psp} is not possible in this direct reflection mode. With the Kretschmann prism coupling (**b**), the wave vector of the light is increased in the prism

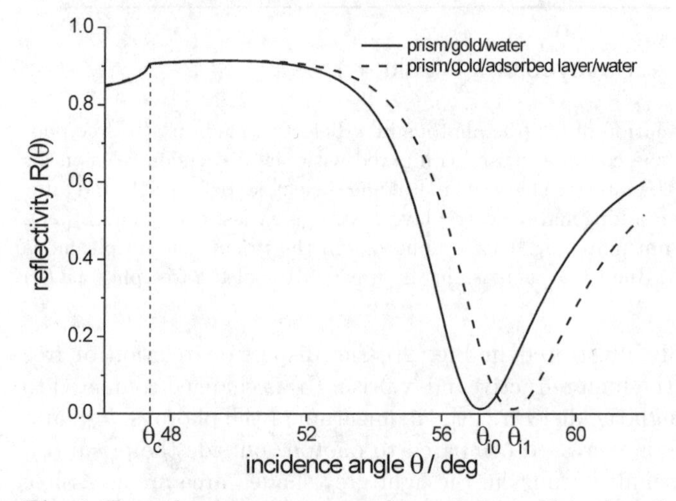

Fig. 4. Typical resonance curves of two systems. Note that the resonance angle of the system is shifted if an additional layer at the gold/water interface is taken into consideration

resonance minimum is observed by plotting the reflectivity, R, versus the applied incident angle. Such a typical resonance curve of the prism–gold–water system is given in Fig. 4.

Starting at small incidence angles, the reflectivity increases until a maximum value is reached at the total reflection edge, θ_c. The reflectivity before that angle is rather high because the thin metal film acts as a mirror that reflects most of the incoming laser light. Further increase of the incident angle increases the momentum of the incoming photons until the resonance condition is satisfied. This results in a decrease in measured reflectivity until minimum reflectivity is reached at θ_{i0}. The resonance dip exhibits a distinct half-width which is dependent on the extent of damping processes in the metal. The resonance angle at which maximal energy transfer occurs is dependent on the dielectric constants of all components and can be derived from the dispersion relation (7) and (2) and (10), respectively:

$$\theta = \arcsin \sqrt{\frac{\varepsilon_{\text{metal}}\varepsilon_{\text{dielectric}}}{(\varepsilon_{\text{metal}} + \varepsilon_{\text{dielectric}})\varepsilon_{\text{prism}}}} . \tag{11}$$

2.3 Surface Plasmons for the Characterization of Thin Layers

Since the evanescent field of the surface plasmons decays exponentially into the dielectric medium, changes in the optical properties of the adjacent dielectric will alter the excitation conditions for plasmons. Therefore, processes near the metal surface like the adsorption of additional dielectric layers can be sensed by the plasmon field and monitored via the resulting changes in the resonance angle of the system.

We consider the prism–gold–dielectric layer system discussed in the previous paragraph and add another layer of dielectric material onto the gold. Here the dielectric constant of these additional layers is assumed to be higher than that of the surrounding medium. Then the effective refractive index of the region sensed by the evanescent plasmon field is increased. As a result the dispersion relation (P2) is shifted towards larger wave vectors as depicted in Fig. 2:

$$k_{p2} = k_{p1} + \Delta k , \tag{12}$$

where k_{p1} and k_{p2} are the wave vectors of the system before and after adsorption of the adlayer. It is important to note that the increment Δk depends on both the refractive index and the thickness of the adsorbed layer.

At a given frequency ω_L the intersection with the dispersion relation of the photons (cf. (3) in Fig. 2) is therefore shifted towards higher k values. Consequently, the corresponding incidence angle has to be experimentally adjusted to achieve resonance again. This situation is shown in Fig. 4, where the new resonance angle θ_{i1} of the associated resonance curve (dashed) can be found at higher angles. This resonance minimum shift is of fundamental importance for experimental application in methods like surface plasmon spectroscopy or microscopy. In nonadsorbing layers ($\varepsilon'' \approx 0$) the measured shift $\Delta\theta$ is proportional to the change in the optical layer thickness Δd and its contrast with the environment, i.e., the difference Δn between the film refractive index and that of the surrounding dielectric medium.

$$\Delta\theta \propto \Delta n \cdot \Delta d . \tag{13}$$

Since an infinite number of combinations of layer thickness and refractive index difference exist that lead to the same resonance shift, one of the parameters has to be known in order to calculate the other. Therefore, at least one of the film parameters has to be assumed or determined by other methods.

2.4 Electromagnetic Field Distribution near the Interface

The electromagnetic field distribution in the layer system can be represented by the magnetic field intensity, H_y^2. The magnetic field intensity found in Fig. 5 was normalized to the incident magnetic field intensity and calculated for a layer system consisting of prism ($\varepsilon = 3.4069$)/gold ($d = 50$ nm, $\varepsilon = -12.1 + \mathrm{i} \cdot 1.3$)/adsorbed layer ($d = 5$ nm, $\varepsilon = 2.25$)/aqueous solution ($\varepsilon = 1.778$). A look at the calculated intensity at incidence angles near the resonance angle (θ_{i1}) reveals a distinct field enhancement in the metal layer, which leads to an increase in the incident field by a factor of about 16 at the metal surface. On both sides of that interface the field intensity decays exponentially, resulting in a certain penetration depth into the dielectric medium. This penetration depth is of the order of the laser wavelength. It has importance for the surface sensitivity of plasmon spectroscopy, because only processes in the evanescent plasmon field are detectable.

In the following, the field intensities at the surface of different metals and the pure dielectric case of a glass/dielectric interface are considered. The calculated intensities are normalized to the incident intensity and are plotted in Fig. 6 together with the corresponding reflectivities as a function of the incidence angle θ^2.

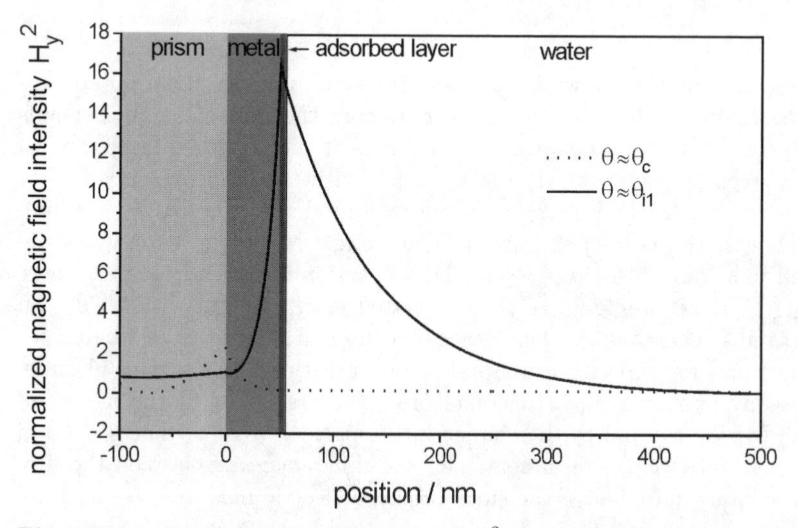

Fig. 5. Normalized magnetic field intensity H_y^2 for a layer system prism/metal/adsorbed layer/water. The exponential decay of the evanescent field into the dielectric is evident. Note that at an incidence angle near the resonance minimum for this system, θ_{i1}, a distinct field enhancement is achieved at the metal surface

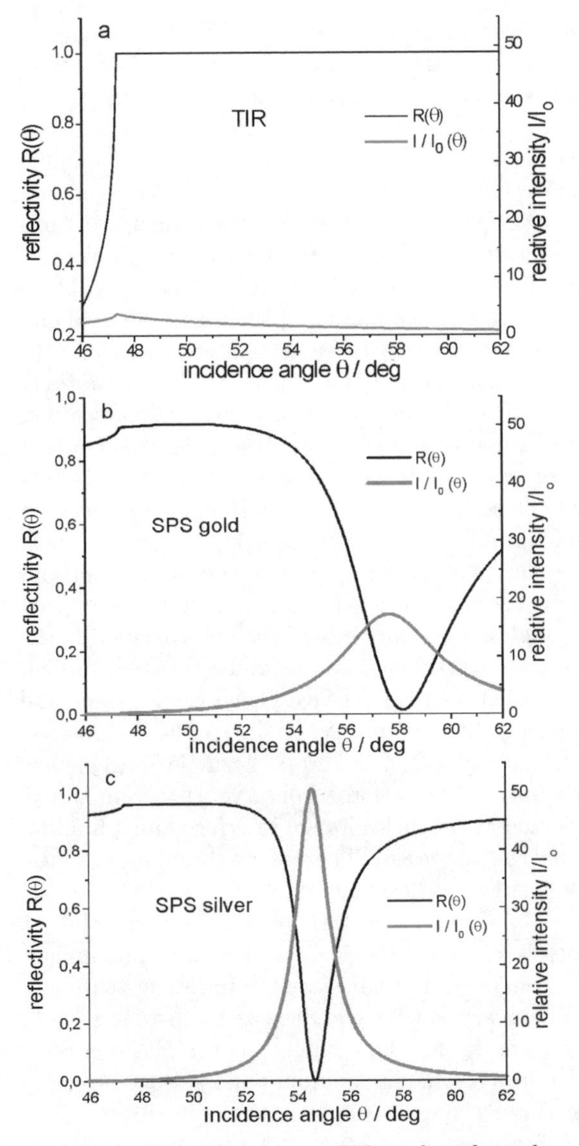

Fig. 6. Comparison between TIR and surface plasmon excitation by considering calculated reflectivities R and relative optical intensities I/I_0 at the surface in the case of (**a**) glass/water interface (**b**) glass/gold/water and (**c**) glass/silver/water

In the case of the glass/dielectric interface (a), the total reflection edge is observed in the reflectivity at θ_c. For angles larger than θ_c, the reflected intensity remains at unity. The field intensity at the surface for TIR (total internal reflection) is plotted, and an increase of a factor of 4 is found at the critical angle. This originates from the constructive interference of the amplitudes of the incoming and reflected electromagnetic fields. This moderate enhancement together with the evanescent character of the light at angles $\theta > \theta_c$ forms the basis for surface sensitive fluorescence spectroscopy [9].

If the glass surface is coated with a thin gold film (b), the total reflection edge is still observable. However, the additional resonance minimum suggests the surface plasmon excitation. Close to the reflection minimum a substantial field enhancement can be found. An enhancement factor of 16 is obtained compared to the intensity of the incident electromagnetic field. If silver is used, an even higher factor of about 50 can be obtained (c). These factors, as well as the half-widths of the resonances, depend on the real and imaginary parts of the dielectric constants. The silver film was modeled with ($\varepsilon = -12.1 + \mathrm{i} \cdot 1.3$) [10], i.e., with an imaginary part of ε'' substantially smaller than that of gold, which results in a much narrower resonance dip. Furthermore, the smaller the value of ε'', the lower the absorption of the sample and the dissipation of optical intensity in the metal, which results in high enhancement factors. This phenomenon is already known from other spectroscopies like surface-enhanced Raman scattering [11].

A closer look at the angular positions of the resonance minimum and the maximum surface intensity reveals that the intensity maximum is shifted slightly to smaller incidence angles. This can be explained by considering the system as a resonator excited by the incoming laser beam. If we sweep through resonance, the phase difference between the surface plasmon mode and the driving photon field behaves in a distinctive way: in a hypothetical lossless metal the phase would vary from 0° to 180° as we cross the angle of resonance, with a sharp step at the resonance. In reflection we observe a coherent superposition of partial waves reflected at the metal/prism interface and a fraction of the surface mode reradiated via the prism. The observed minimum in the resonance scan can now be explained as destructive interference between the two partial waves, which are phase-shifted by 180° relative to each other at the reflection maximum. Therefore, no shift between resonance minimum and intensity maximum would be seen in such a lossless metal. In real systems, any losses, as described by the imaginary part of the dielectric constant $\varepsilon'' > 0$, slightly shift and considerably smear the phase change at the plasmon resonance angle, and consequently the resonance minimum and surface intensity peak drift apart. Since the damping in gold is stronger than in silver ($\varepsilon''_{\mathrm{Au}} > \varepsilon''_{\mathrm{Ag}}$), the difference between these angular positions is larger than for the silver system depicted in Fig. 6 [2]. Silver is frequently used as the metallic layer because of the strong enhancement of the surface mode and the narrow resonance dip. Due to the known stability of the inert gold surface against buffer solutions and oxidizing agents, the latter is preferred in our experiments.

2.5 Fluorescent Chromophores near Metal Surfaces

Consider a fluorophore which is excited either by direct illumination or by the evanescent field of a surface plasmon in front of a planar metallic surface. Since the metal film serves as a mirror, the reflected field interferes with the emitting dipole. If the reflected field is in phase with the dipole oscillations, it will be excited by the reflected electromagnetic wave. The dipole will be driven harder and consequently the emission will be enhanced. If the reflected field is out of phase, the emission will be reduced. Thus, the dipole can be considered a forced, damped, dipole oscillator [12]. With increasing distance between the dipole and the metal surface the phase difference between the incident and reflected light changes, which results in an oscillating emission rate of the dipole. Furthermore, with increasing distance of the dye from the metal, the strength of the oscillation will decrease. The radiation field of the dipole at the surface weakens with increasing distance from the surface and thus the strength of the reflected field will also decrease. These predicted features were verified by measuring the lifetime of Eu^{3+} ions in front of a Ag mirror [13,14]. In addition to these features, strong quenching of the fluorescence light was found for small emitter–surface separations. This phenomenon cannot be explained by simple interference and is attributed to direct coupling between the dipole field and the surface plasmon modes.

Decay Channels for Excited Chromophores near Metal Surfaces

Mainly three different decay channels for fluorescence near the surface can be considered, as schematically presented in the cartoon in Fig. 7.

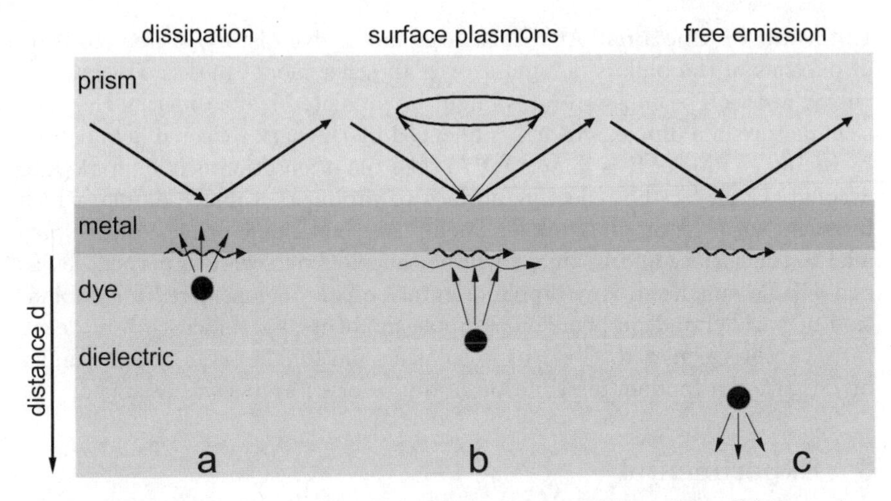

Fig. 7. Major decay channels for fluorescence near metallic surfaces. While at close distances quenching due to the metal layer is found (**a**), the excitation of plasmons by the red-shifted fluorescence light dominates the fluorescence intensity up to a distance of about 20 nm (**b**). This new PSP mode reradiates via the prism in a cone at angles corresponding to the dispersion relation at the Stokes-shifted wavelength. Finally, at large distances from the metal the free emission of photons can be found (**c**)

Nonradiative Transition and Exciton Coupling. Very close to the metal surface, up to 10 nm from the surface, electron-hole pairs (excitons) can be excited by the near field of the dipole, and thus the interaction is dipole–dipole. For very small metal–dye distances, d, the dipole field is dominated by the near field, the strength of which decreases as d^{-3}. The classical Förster model for the interaction between two individual dipoles involves the distance dependence of both acceptor and donor, and thus results in a d^{-6} dependence. However, integrating over all possible sites of an entire plane of acceptor sites will yield a d^{-4} dependence. The transferred energy is dissipated in the metal and thus lost for fluorescence detection.

Coupling to Surface Plasmon Modes. Consider a dipole located a few nanometers away from the surface, up to 20 nm. An excited dipole close to a metal surface can excite surface plasmon modes at its Stokes-shifted emission wavelength. This again can couple to light if momentum matching via a prism occurs [15]. This type of excitation typically becomes dominant at dye–metal distances of about 20 nm [16]. The angular distribution of this plasmon-coupled fluorescence emission is determined by the fluorophore emission spectrum and the surface plasmon dispersion [12]. The back radiation through the prism will lead to an emission cone [17], since the wave vector of the excited plasmons is fixed, but no plane of incidence can be defined any more because the fluorophores emit in random directions.

Emission of Photons. At rather large metal–dye distances, free emission of photons in the bulk is dominant over all other modes of deexcitation discussed above. The emission rates might be predicted by considering the classical picture of a dipole, which is influenced by the back reflected field caused by its image dipole. It is important to take the orientation of the dipole into account. Consider the reflecting surface to produce an image dipole, which interacts with the original dipole. Then it can be assumed that a dipole parallel to the surface and its image will be cancelled out, while a perpendicular one will be enhanced. Any dipole orientation can be considered a combination of parallel and perpendicular dipole moments, for which certain decay rates can be calculated. For dyes in solution and for the layer systems under investigation a random distribution of dipole orientations was assumed.

3 Experimental

The setup is a common surface plasmon spectrometer [1] which was modified with a fluorescence detection units [2]. As schematically depicted in Fig. 8, a HeNe laser ($\lambda = 632.8$ nm) and a frequency-doubled Nd:YAG laser ($\lambda = 532$ nm) were used. The excitation beam passes two polarizers, by which the intensity of the incident light and its TM polarization can be adjusted. Using

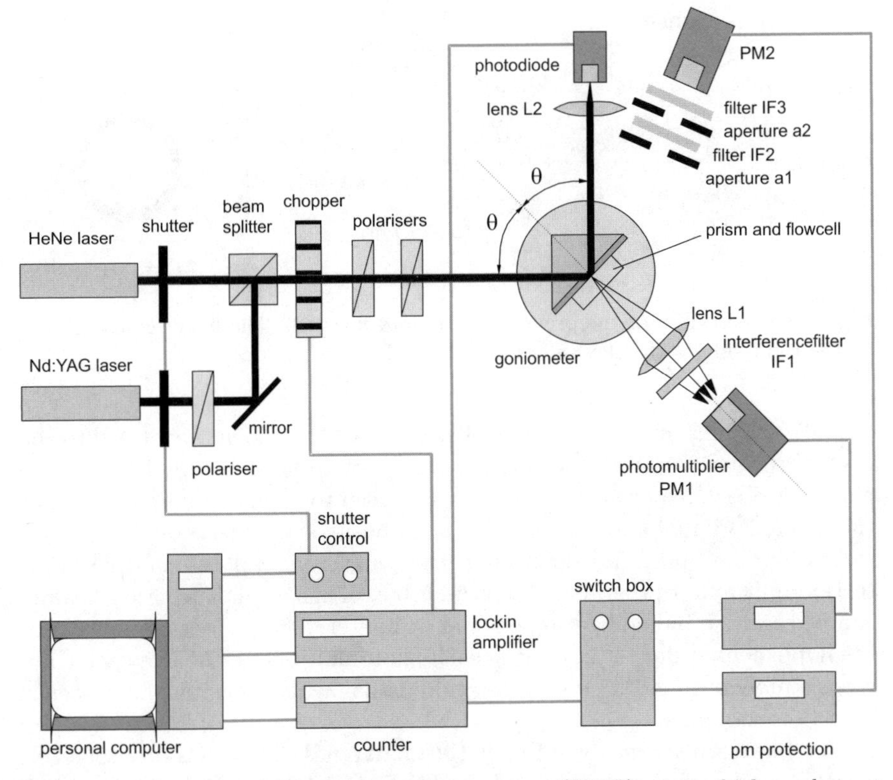

Fig. 8. Surface Plasmon Fluorescence Spectroscopy (SPFS) setup which can be run with two photomultipliers and at two different laser wavelengths

a beamsplitter and two programmable shutters, the incident wavelength can be easily changed by blocking one of the laser beams and passing the other on to the sample.

The incident laser is reflected off the base of the coupling prism (90°, LaSFN9) and the reflected intensity is focused by a lens (L2, $f = 50$ mm) for detection by a photodiode. In order to allow for noise-reduced and daylight-independent measurements of the reflected intensity, the photodiode is connected to a lock-in amplifier. This unit filters out all frequencies that are not modulated at the operating frequency of the attached chopper.

The sample is mounted on a two-phase goniometer which can be moved in $\Delta\theta = 0.001°$ steps by the connected personal computer. According to the reflection law the angular position of the optical arm holding the detection unit (detector motor) is adjusted during the measurements. The sample is mounted on an xy-stage and a tilting table, which allow for optimal adjustment of the sample.

In order to detect fluorescence emission from the sample, a collecting lens (L1, $f = 50$ mm) focuses the emitted light though an interference filter

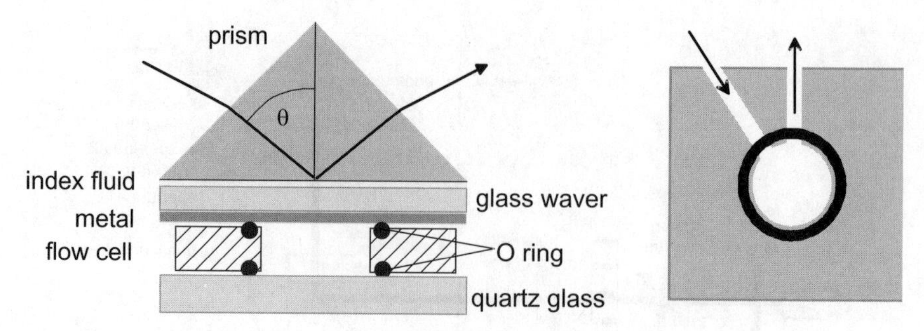

Fig. 9. Cross-section of the prism, sample, and flow cell. The latter is also shown on the right side, with inlet and outlet channels

($\lambda = 670 \pm 2$ nm) into a photomultiplier tube (PM1), which is attached to the back of the sample. Additionally, a second photomultiplier tube PM2 can be placed on the detector arm. This latter was used to monitor fluorescence light that was reradiated and emitted via the opposite side of the coupling prism. Both photomultipliers are connected to a counter via a photomultiplier protection unit and a programmable switch box. Thus, signals from both units can be recorded online by the personal computer. The protection unit closes the implemented shutter in front of each photomultiplier if the irradiation exceeds a predefined level, in order to avoid damaging the sensitive fluorescence detection equipment.

As shown schematically in Fig. 9, the quartz glass flow cell is positioned on a low-fluorescence quartz glass slide and sealed with Viton O-rings. The glass cover is placed on top of the flow cell, while the evaporated metal film points towards the cell. Finally, a high refractive index prism (LaSFN9, $n = 1.845$ at $\lambda = 632.8$ nm) is mounted on top of the glass sample. In order to allow for optimal coupling of incident light to the plasmon modes of the metal, a thin film of refractive index matching oil is added between the two glass units. The flow cell is equipped with an inlet and outlet and can hold volumes of up to about 90 μl. For the injection of analyte samples one-way plastic syringes are used, but to rinse the cell with pure buffer and to rinse the sample after adsorption processes a peristaltic pump is connected via Teflon tubing.

4 Results and Discussion

4.1 Experimental Verifcation of Surface Field Enhancement

In Fig. 10, the fluorescence signal under TIR and SPS conditions is given for a random distribution of fluorophores (MR121, structural formula also given in Fig. 10) in a 20 nm thick polyvinyl alcohol (PVA) film. In order to ensure identical experimental conditions, half of a high refractive index LaSFN9 wafer was covered with a 50-nm thin silver film, such that TIR and SPS experiments could be performed on the same sample. After spin coating

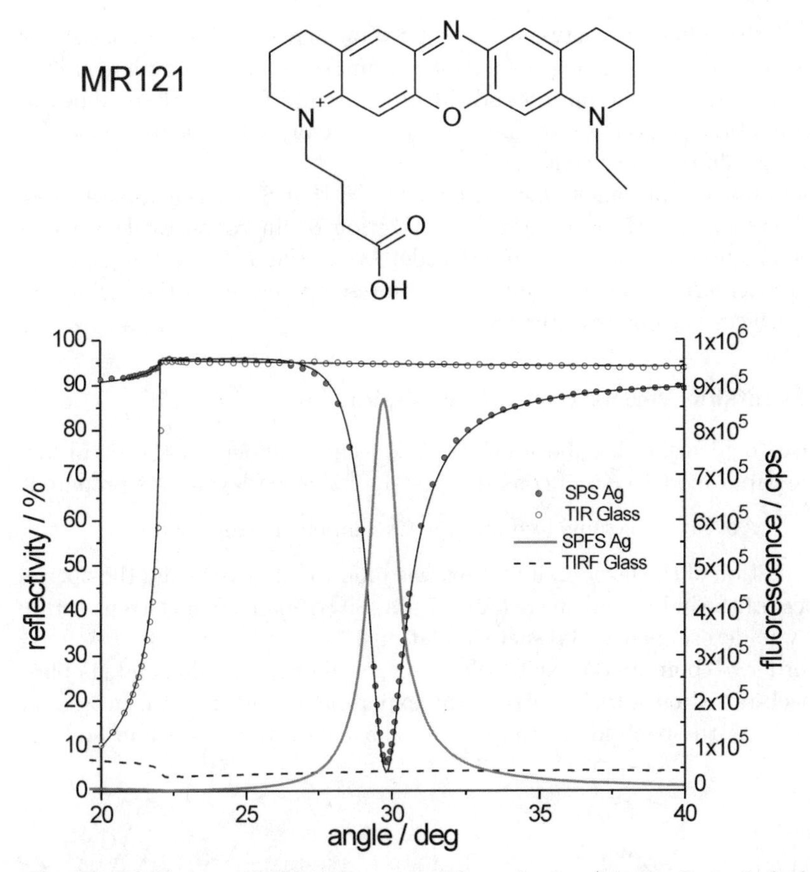

Fig. 10. SPFS scan of MR121 (structural formula given) in a polyvinyl alcohol film spin-coated on silver and LaSFN9 glass (with Fresnel fits for the reflectivity). Due to plasmon excitation, a clear fluorescence maximum is seen near the resonance minimum at $\theta = 29.8°$, while under TIR conditions (glass) no such maximum was found for angles greater than the critical angle

the MR121/PVA film, SPFS scan curves of both areas were recorded as described before using p-polarized HeNe laser light with $\lambda = 632.8$ nm. While for TIRF (total internal reflection fluorescence) only background fluorescence was observed, strong signal enhancement was evident in the SPRF scan close to the resonance minimum for plasmon excitation at $\theta = 29.8°$. Note that the angular dependence of the sharp fluorescence peak was found to follow the calculated field distribution, as discussed in Fig. 6. If the polarization of the incident electromagnetic waves was changed from p- to s-polarization (not shown), no surface plasmons were excited any more, and the fluorescence signal did not exceed the background level at any angle.

Another disadvantage of TIRF with prism coupling is that strong background fluorescence can result from intrinsic fluorescence of the LaSFN9 prism. If the difference between the excitation and emission wavelength of the dye is small, the exciting laser light can interfere with the Stokes-shifted fluorescence, due to the finite bandwidth of the interference filters. The additional metal film in SPFS samples acts as a mirror and helps to reduce this background fluorescence contribution.

Preliminary experiments have demonstrated that SPFS is a valuable tool that allows for a surface-sensitive investigation of fluorophores. Due to its various advantages over common TIRF detection, the SPFS technique is the method of choice if the environment of fluorescent species in the vicinity of metal surfaces is to be investigated.

4.2 Frontside Versus Backside Emission

In order to identify the above-mentioned decay channels in the employed prism setup a model system consisting of the following layers was prepared:

Au/SiOx/biotinylated silanes/Cy5-labelled streptavidin

The formation of the layer architecture was monitored by SPS and the optical thickness of the individual layers (Au: 47 nm, SiOx-Silane: 9 nm, streptavidin: 5 nm) was determined by Fresnel simulations.

In order to compare the excited fluorescence via free emission and via plasmon back-radiation into the prism, the experimental setup given in Fig. 11 was used. At the backside of the prism, adjacent to the metal film and the

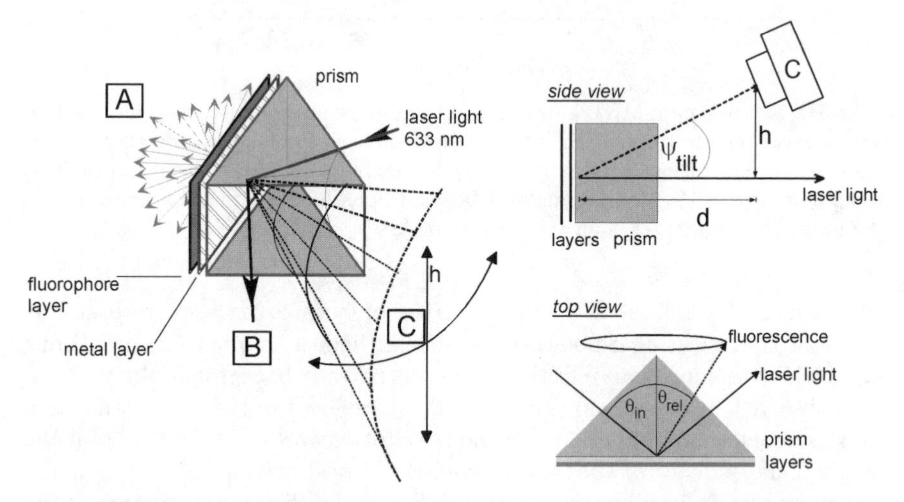

Fig. 11. Schematic illustration of the prim setup. The reflected laser light is monitored by detector B. Photomultiplier A detects the randomly directed free emission of photons from the backside of the sample, while unit C is used to sample the emission cone of the plasmon-coupled fluorescence at the opposite side of the prism. The position of photomultiplier C is determined by θ_{rel}, d and h, as defined in the top and side view of the setup

dye layer, a lens (not shown) was mounted to collect the radiated free emission into photomultiplier A. Simultaneously, the intensity of the reflected laser beam was recorded by photodiode B in order to record the SPS signal. Additionally, the emission cone of the plasmon-coupled fluorescence on the front side of the prism was sampled by photomultiplier C. The position of the latter was determined by the emission angle θ_{rel}, the distance to the prism d, and the distance to the plane of incidence h as defined in Fig. 11. Both photomultipliers were equipped with an interference filter ($\lambda = 670$ nm) in order to block all wavelengths different from the emission of the Cy5 dye.

Free Emission of Photons

The SPFS scan curves as measured by detectors A and B before and after adsorption of Cy5-labelled streptavidin (SA-Cy5) on the silane functionalized surface are presented in Fig. 12. Due to the adsorption of the molecules on the surface the resonance minimum was shifted to higher incidence angles. Compared to the low fluorescence background a strong signal enhancement was observed near the resonance minimum after immobilization of the labeled proteins. Two important features need to be emphasized here: although for the measurements on silver in Fig. 10 another layer architecture was investigated, comparison of systems with silver and gold metal layer reveals a much broader peak in the case of gold, originating from the stronger damping of the electromagnetic wave. Again, the maximum of the fluorescence can be

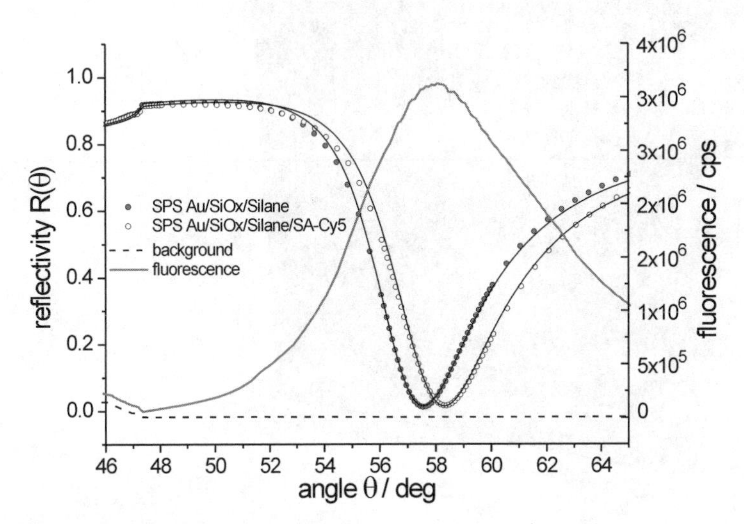

Fig. 12. SPFS scan (free emission of photons) of the layer system Au/SiOX/Silane/biotin/SA-Cy5 before (– – –) and after (—) adsorption of the Cy5 labelled streptavidin onto the silane functionalized surface as measured with detector A in Fig. 11

detected close to the reflection minimum of the surface plasmon for HeNe laser light but slightly shifted to smaller incidence angles.

Different from the experiment in Fig. 10, here the SiOx and the protein layer guarantee a minimum metal–dye distance of not less than 14 nm. Hence, the rather high fluorescence signal of the free emission at the base monitored by detector A proves that quenching processes are already largely reduced at this distance.

Surface Plasmon Modes

The emission characteristics of the plasmon-coupled fluorescence process was monitored by increasing the angle of incidence θ_{in} from 46° to 64° stepwise ($\Delta\theta_{in} = 1°$) and performing scans with detector C along the "frontside" of the prism ($\theta_{rel} = 15 - 85°$) at each incidence angle. The position of the photomultiplier was given by the angle Ψ_{tilt}, which can be expressed in terms of the distances $h = 4$ cm and $d = 10$ cm. The angular resolution was determined by the pinhole (2 mm) in front of detector C and was calculated to be 0.76°. Figure 13 summarizes the scans obtained and reveals that depending on both the excitation angle θ_{in} and the emission angle θ_{rel}, an intensity maximum can be found for the plasmon-coupled fluorescence.

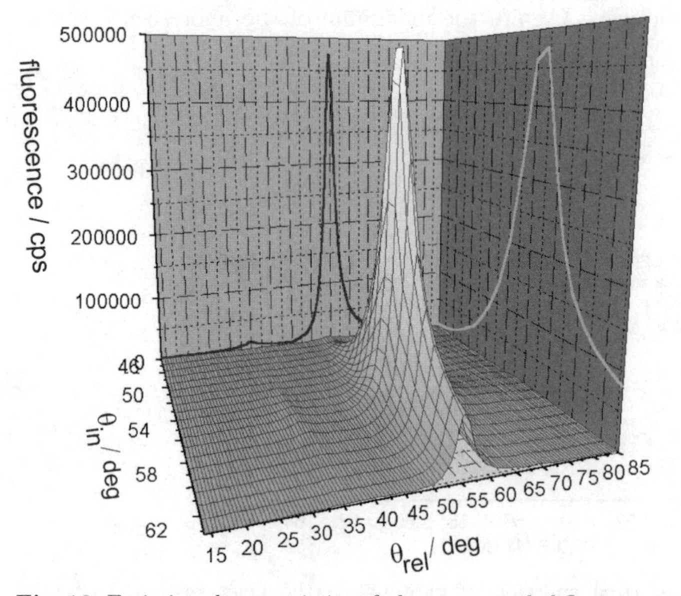

Fig. 13. Emission characteristics of plasmon-coupled fluorescence depending on the incident angle $\theta_{in} = 46 - 64°$ and the emission angle $\theta_{rel} = 15 - 85°$ as measured by detector C (cf. Fig. 11). Maximum fluorescence can be found at $\theta_{in} = 58°$ and $\theta_{rel} = 55°$

The maximum at $\theta_{\text{in}} = 58°$ was already observed in Fig. 12 for the free emission of photons, and results from sweeping through the resonance of the layer system. The more the plasmon resonance is excited, the higher the field at the surface. Therefore, a fluorescence maximum is found close to the plasmon resonance minimum of $\theta_{\text{in}} = 58°$. Scaling the fluorescence peak in Fig. 12 (detector A) and comparing the result with the peak at $\theta_{\text{in}} = 58°$ in Fig. 13 revealed identical curve shapes (not shown).

The maximum at $\theta_{\text{rel}} = 55°$ corresponds to the detector crossing the emission cone sketched in Fig. 11, and proves that the plasmon-coupled fluorescence light was emitted into a narrow angular range. Due to the red shift of the fluorescence light from $\lambda_{\text{ex}} = 632.8$ nm to $\lambda_{\text{em}} = 670$ nm, the fluorescence maximum and the corresponding resonance minimum at this wavelength appears at smaller angles, as measured for the exciting HeNe laser light ($\theta_{\text{rel}} = 58°$). According to the dispersion relation of surface plasmon, light of $\lambda = 670$ nm excites a plasmon mode at a different incidence angle than light of $\lambda = 632.8$ nm. The small peak around $\theta_{\text{rel}} = 35°$ is considered to be caused by multiple reflections of the fluorescence light in the prism (symmetric to $\theta = 45°$) and can be neglected.

In order to sample the shape of the emission cone, more scans with detector C were performed at different heights $h = 2$, 4, and 6 cm, respectively, while the distance $d = 15$ cm was kept constant. An increase in the distance h between detector C and the plane of incidence causes the photomultiplier to cross the emission earlier. This results in a shift of the maximum fluorescence to smaller angles θ_{rel} for increasing heights h (Fig. 14). Note, that the intensity variation in this set resulted from experimental difficulties in aligning the detector and is of no importance here. However, if the angles of maximum intensity were converted to the x-component of the surface plasmon wavevector, $k_{\text{psp},x}$, all three maxima appeared to be superimposed and thus proved the measured fluorescence to be identical in its origin.

The intensity of the fluorescence measured for free emission and for the plasmon-coupled case cannot be compared directly, because only a cross section of the emission cone was recorded. Due to experimental uncertainties based on the difficulties in aligning detector C, a simple evaluation of the total back-radiated emission by integrating over the complete cone is not possible. However, the absolute intensities were not under investigation; rather, the presence of the individual decay channels was to be verified for the prism setup employed.

Although the measurement of back-coupled fluorescence has been employed by other researchers for sensing purposes [18], detection of the free emission in SPFS as described here is a much simpler experimental approach. Additionally, extension to fluorescence microscopy is possible only by employing measurements of the fluorescence from the backside of the prism.

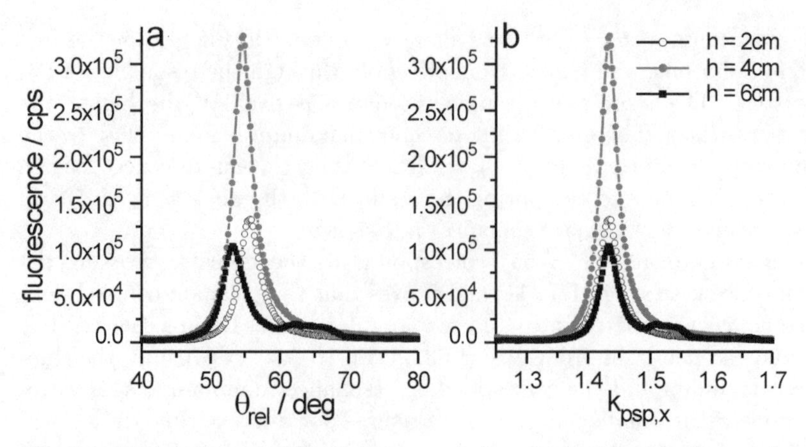

Fig. 14. Plasmon-coupled fluorescence emission at a fixed excitation angle of $\theta_{\mathrm{in}} = 56^{\circ}$. Scans around the prism with detector C revealed fluorescence emission in a narrow range of angles θ_{rel}. (**a**) Scans at increasing distances $h = 2$, 4, and 6 cm relative to the plane of incidence (and a detector–prism distance of 15 cm) revealed a peak shift to smaller angles, corresponding to earlier crossing of the detector through the emission cone. (**b**) Conversion of θ_{rel} into internal wave vectors proved that all emission maxima correspond to the same k_{psp}

5 Conclusions

We have shown that fluorescence dyes can be excited by the evanescent surface plasmon field close to metallic surfaces, and that the enhancement factor is strongly dependent on the optical properties of the metal. For maximum fluorescence emission it is necessary to choose the optimal spacing between the dye and the metal surface. For small distances between dye and metal the exponentially decaying evanescent field exhibits a maximum, and so does the probability of exciting the fluorophore by the surface field. On the other hand, proximity to a metal surface gives rise to quenching effects which reduce the measurable fluorescence intensity.

We have shown that both the free emission of photons and plasmon-coupled fluorescence can be measured using the prism setup shown before. Compared to the experimental comfort of detecting fluorescence from the backside of the prism (detector A in Fig. 11), more effort is required to align the spatial and angular position of detector C. Since no clear advantage of measuring the plasmon-coupled fluorescence was seen, common SPFS measurements using detector A are preferred for further experiments. Additionally, extension of the existing setup to fluorescence microscopy is (easily) possible only from the position of detector A at the prism setup. Measuring plasmon-coupled fluorescence might be useful for special applications which make use of the distance dependent quenching profile, but this was not found to be a practical alternative to free photon detection.

Acknowledgements

We are grateful to Dr. R. Herrmann and Dr. P. Sluka (Roche Diagnostics) for providing the MR121 fluorescence dye, as well as thiolated streptavidin. This work was partially supported by two EU grants: BMH4-CT96-1081 and QLK1-2000-01658.

References

1. W. Knoll: Annual Review of Physical Chemistry **49**, 569 (1998)
2. T. Liebermann, W. Knoll: Colloids and Surfaces A: Physicochemical and Engineering Aspects **171**, 115 (2000)
3. H. Raether: *Surface Plasmon on Smooth and Rough Surfaces and on Gratings*, Springer Tracts in Mod. Phys. **111** (Springer, Berlin, 1988)
4. V.M. Agranovich: *Surface Polaritons* (North-Holland, Amsterdam, 1982)
5. P. Yeh: *Optical Waves in Layereed Media* (John Wiley & Sons, New York, 1988)
6. E. Kretschmann: Zeitschrift für Physik **241**, 313 (1971) 241.
7. E. Burstein, W.P. Chen, A. Hartstein: Journal of Vacuum Soc. **11**, 1004 (1974)
8. J.R. Sambles, G.W. Bradberry, F. Yang: Contemporary Physics **32**, 173 (1991)
9. D. Axelrod, T.P. Burghardt, N.L. Thompson: Biophys. Bioeng. **13**, 247 (1984)
10. P.B. Johnson, R.W. Christey: Phys. Rev. B **6**, 4370 (1972)
11. A. Nemetz, W. Knoll: Raman Spectrosc. **27**, 587 (1996)
12. R. Amos, W.L. Barnes: Phys. Rev. B. **55**, 7249 (1997)
13. H. Kuhn: J. Chem. Phys. **53**, 101 (1970)
14. W. Weber, C.F. Eagon: Opt. Lett. **4**, 236 (1979)
15. I. Pockrand, A. Brilliante, D. Möbius: Chem. Phys. Lett. **69**, 499 (1980)
16. M. Kreiter: Oberflächenplasmonen-artige Resonanzen auf metallischen Gittern. Ph.D. Thesis, University of Mainz, Germany (2000)
17. R.E. Benner, R. Dornhaus, R.K. Chang: Opt. Commun. **30**, 145 (1979)
18. J.W. Attridge et al.: Biosens. Bioelectron. **6**, 201 (1991)

Optical Characterization of In(Ga)As/GaAs Self-assembled Quantum Dots Using Near-Field Spectroscopy

Y. Toda and Y. Arakawa

1 Introduction

Reduced dimensions of carrier motion down to zero-dimensional (0D) can be realized in semiconductor quantum dots. In such 0D systems, fully quantized states will significantly improve performance in photonic and electronic device applications [1]. One of the most representative structures is self-assembled quantum dots (SAQDs), using the Stranski–Krastanow (SK) growth mode, in which dots form spontaneously during the overgrowth of highly mismatched materials, as shown in Fig. 1. Because of the absence of processing by lithography and etching, SAQDs provide defect-free structures with high emission efficiency. In addition, their high dot density ($> 10^{11}$ dots/cm^2) and ability to stack vertically enhance carrier density, which is important for realizing quantum dot lasers with low threshold current. Indeed, injection lasers with low threshold current have already been demonstrated using InGaAs/GaAs SAQDs [2].

Another fascinating property of SAQDs is the size of the dots. In SAQDs, strong confinement on length scales down to 10 nm in all three directions can be realized (Fig. 1b), and will produce large sublevel separations in their density of states (DOS). In such DOS comparable to an artificial atom, optical spectroscopy enables us to investigate the 0D nature of excited carriers. For example, optical excitation can probe the relaxation mechanism of excited carriers, which is an important subject of discussion not only because of device applications but for understanding the fundamental physics of fully quantized 0D systems. In this chapter, we consider the optical properties of $In_xGa_{1-x}As$ SAQDs characterized by various types of optical spectroscopy such as photoluminescence (PL), PL excitation (PLE), and Raman scattering. In addition to conventional far-field spectroscopy, we perform advanced near-field optical measurements. It has been shown that PL linewidths of single SAQDs are orders of magnitude narrower than observed in ensemble spectra [3]. Due to the reduction of the inhomogeneous ambiguity, near-field optical probing of single quantum dots is thus a powerful tool for providing precise information about such 0D systems. Furthermore, near-field imaging allows sensitive measurements of local environments around individual SAQDs. This observation also applies to the fundamental physics of a 0D system.

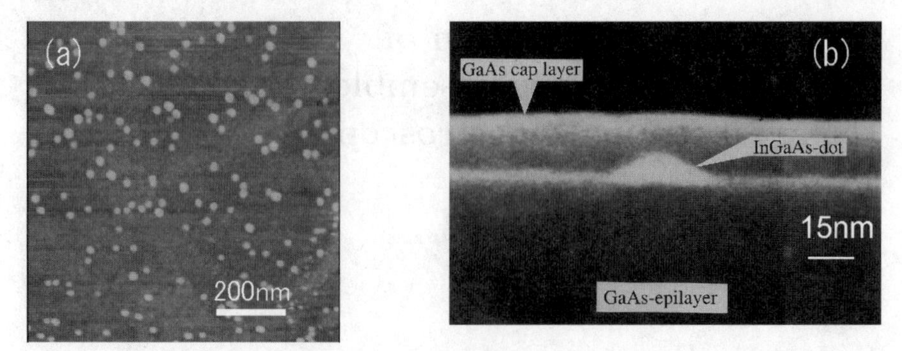

Fig. 1. InGaAs/GaAs self-assembled quantum dots. (**a**) [001] plan-view atomic-force microscope (AFM) image of uncapped sample. (**b**) Cross-sectional scanning electron microscope (SEM) image of the sample

In Sect. 2, we briefly review possible relaxation mechanisms in quantum dots. Section 3 presents several experimental results on far-field measurements and briefly summarizes the ensemble optical properties of SAQDs. In Sect. 4, we show some experimental results from single quantum dots using near-field optical spectroscopy. We present characteristic 0D-like behavior of ground state emission under various conditions. The results enable us to discuss dephasing processes and confinement effects in SAQDs. Comparing between various near-field spectroscopic results, we consider the relaxation process and electron–phonon interactions in SAQDs. In order to evaluate the feasibility of advanced coherent engineering of carrier wavefunctions, we demonstrate near-field coherent spectroscopy of single SAQDs.

Most of the experimental work presented here is focused on a single sample of $In_xGa_{1-x}As/GaAs$ SAQDs so as to compare the different experimental results [4]. The SAQDs have average base length $\sim$ 20 nm, height $\sim$ 7 nm, and density $> 10^{10}/cm^2$, from AFM study of an uncapped reference sample (see Fig. 1a). The In content, x is estimated to be more than 0.5. In order to shift the PL peak to shorter wavelengths where optical detection with high quantum efficiency is available, the sample was annealed at 700°C after growing a 140-nm thick GaAs cap. We believe that this annealing only changes the luminescence energy and does not affect the intrinsic optical properties.

2 Relaxation Mechanism

Photoexcited carriers will dissipate energy and return to their initial thermal equilibrium. This process is called energy relaxation. In semiconductors, one efficient relaxation process is inelastic scattering with phonons. Because in quantum dots the confinement on length scales down to 10 nm in all three directions produces sublevel separations greater than the phonon energy, relaxation between the discrete levels will be slow unless the energy level separation equals the longitudinal optical (LO) phonon energy. In addition, relaxation through LO phonon scattering is expected to be very restricted due to the phonons' small energy dispersion. Therefore, a markedly reduced energy relaxation rate, the so-called "phonon bottleneck," is predicted in the discrete states of quantum dots [5]. Although theoretical discussions of the phonon bottleneck should be correct, this effect seems unlikely to be present in SAQDs, since optical measurements show intense luminescence due to ground-state recombination. Several alternative relaxation mechanisms have also been discussed.

The Auger process is the candidate for fast relaxation at high carrier density [6]. Consider two carriers in a dot excited state. In the Auger process, one carrier relaxes to lower states and transfers its energy to another carrier. Because continuum sublevels exist in the barrier, one of the carriers can find a final state in the continuum in order to satisfy energy conservation. Therefore a high density of carriers can overcome the phonon bottleneck even when relaxing to discrete energy levels with large energy separations.

In polar semiconductors, the most effective relaxation mechanism is inelastic scattering on LO phonons via Fröhlich coupling. Although the phonon bottleneck indicates that polar LO phonon-related transition is only possible when the energy splitting between the initial and final state exactly matches the phonon energy, optical measurements on SAQD suggest that inelastic scattering on LO phonons still plays an important role in carrier relaxation [7]. For example, PLE spectra of inhomogeneously broadened ensembles show broad multiple LO phonon resonances, as discussed in Sect. 3.3. Similar multiphonon resonances can be observed in near resonance excited PL as well [7].

Many researchers have mentioned such multiphonon relaxation mechanisms, mainly based on experimental results. The two-phonon process has been studied theoretically by Inoshita et al. [8]. Simultaneous scattering with LO and LA phonons can relax the restriction of the LO phonons' small energy dispersion, and thus results in a high phonon scattering rate. Neverthless this model does not provide an explanation for strongly confined dots in which the energy sublevels exceed the LO phonon energy by a factor of two or more. Recent improved theoretical studies using a coupled-mode equation result in a different prediction from the original phonon bottleneck. Li et al. have demonstrated that efficient relaxation can be achieved even when restricted to the intrinsic phonon scattering mechanism [9]. Due to the anharmonic coupling of the LO phonons to the acoustic phonons, the calculated relaxation rate reaches higher than 10^{10} s^{-1} over a wide energy range of tens of meV around the LO phonon energy.

Recent experimental observations have shown that enhanced LO phonon–electron coupling, where LO phonons interact coherently with electrons, can be realized in SAQDs. Using far infrared magneto-spectroscopy, Hameau et al. have observed huge anti-crossings of the electron states due to strong coupling with the LO phonons [10]. Similar enhanced phonon coupling has been reported by several groups [11]. However, the detailed mechanisms of such enhancement are still controversial.

3 Optical Properties of Self-assembled Quantum Dots: Far-Field Analysis

3.1 Photoluminescence Spectroscopy

At present, PL spectroscopy is widely and routinely one of the most valuable techniques for investigating the fundamental optical properties of a sample. In a typical PL experiment, the carriers are created above the lowest energy band gap. Excited carriers relax down to quasi-equilibrium by appropriate scattering events, and then emit photons when radiatively recombining. In quantum dots, the radiative recombinations at low carrier density usually occur in the lowest energy level. Therefore PL of SAQDs provide useful insights into the radiative recombination of carriers in the ground state. Figure 2 shows ensemble PL spectra of InGaAs/GaAs SAQDs. Due to the strong confinement of the SAQD structure, a size distribution leads to large inhomogeneous broadening in ensemble PL spectra. The observed Gaussian profile with a broad linewidth of $\sim$ 80 meV of PL from SAQDs is thus attributed to inhomogeneous broadening. Upon increasing the excitation power, two additional peaks appear in the higher-energy region of the spectra. Interband recombinations in the higher excited states can also emit photons with a wave vector selection rule satisfied. Because only one pair of carriers at a time can exist in the lowest state, interband recombinations will take place

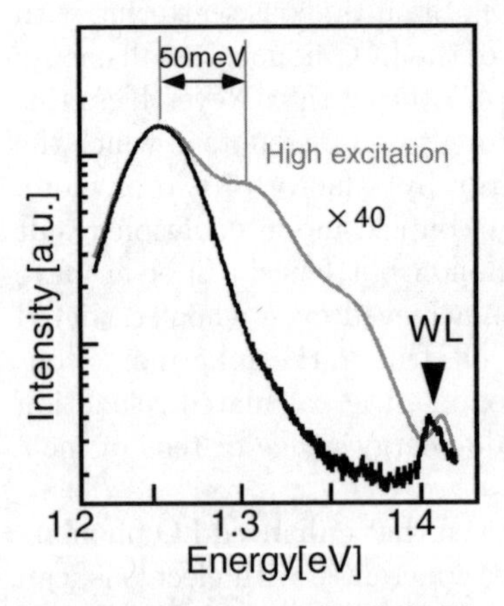

Fig. 2. Nonresonant far-field PL spectra in the sample at 8 K. The solid and grey lines show ensemble PL at low-power excitation $P_{ex} = 5$ W/cm^2 and high-power excitation $P_{ex} = 70$ W/cm^2, respectively, with excitation energy $E_{ex} = 1.959$ eV. WL indicates the emission from the wetting layer

after the ground state fills. The observed PL peak on the higher energy side could thus be attributed to the higher excited states of dots [12]. We used here the SAQDs with post-growth annealing at 700°C. The PL from dots is shifted toward higher energies upon annealing. This blueshift is mainly due to out-diffusion of In atoms. The blueshift can be also seen in PL peaks on the higher energy side. Moreover, the sublevel separations decrease with increasing annealing temperature. The reduction in sublevel splitting might be due to an increased effective size of dots. Although thermal annealing changes the luminescence energy, intrinsic optical properties such as PL and PLE structures appear qualitatively unchanged except for the sample annealed at too high a temperature.

3.2 Magneto-Optical Spectroscopy

The application of a magnetic field reveals the effective carrier confinement in quantum dots. In a weak magnetic field in which the Landau orbit is far larger than the carrier wave function, the magnetic field can be treated as a perturbation to the confinement potential. In this case, the lowest quantized energy displays a quadratic shift to higher energy in a changing magnetic field. This phenomenon is the so-called diamagnetic shift, and can be observed by monitoring the PL peak energy.

Magneto-PL spectra in the Faraday configuration, where the magnetic field is applied along the growth axis ($B \perp Z$), are shown in Fig. 3. In this figure, the PL peak shifts toward higher energy with increasing magnetic field. As mentioned in the previous subsection, far-field PL occurs in inhomogeneous ensembles. Therefore the observed diamagnetic shift may include the

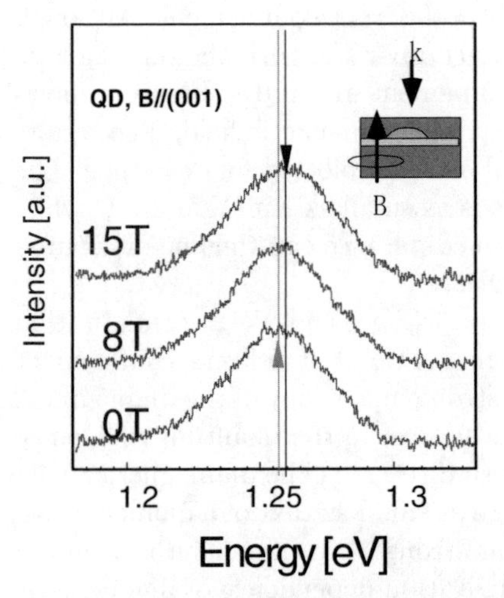

Fig. 3. Magneto-PL spectra of SAQDs in the Faraday configuration. The grey and solid lines are the PL peaks at $B = 0$ T and 15 T, respectively

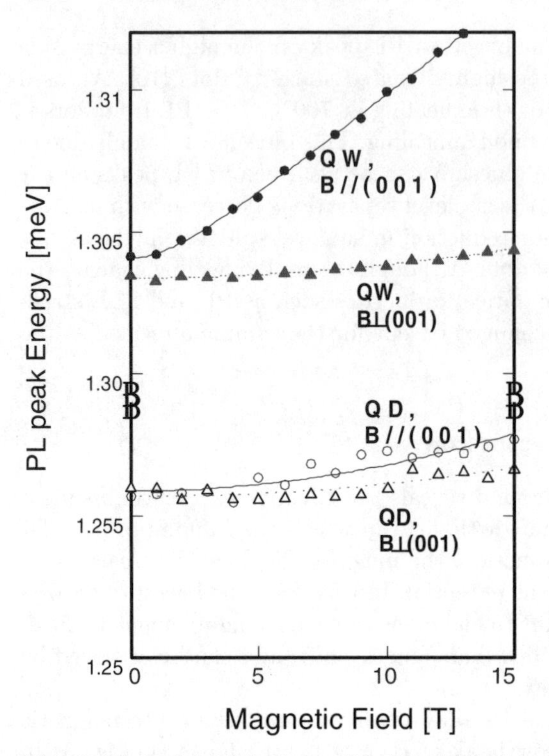

Fig. 4. PL peak positions as a function of the magnetic field. Circles and triangles correspond to the Faraday and Voigt configuration, respectively. The solid lines are least-squares fits to a parabola. Note the break in scale between 1.26 and 1.30 eV

distribution of individual dots with slightly different confinement. Although the PL linewidth of dots is still broad by comparison with the diamagnetic shift, we can roughly evaluate the 3D confinement in SAQDs. Figure 4 shows the energy shifts of PL peaks as a function of the magnetic field. The results for a 140-Å AlGaAs/GaAs quantum well are also plotted for reference. The shift of the PL peak of the quantum dots is as small as 2 meV at 15 T, while that of the quantum wells is 9.3 meV. This difference can thus be attributed to the strong lateral confinement of SAQDs.

In Fig. 4, the open circles indicate the PL peak in Voigt configuration ($B \parallel Z$). In this configuration, the diamagnetic shift reflects confinement along the growth direction. As expected, strong anisotropy of the diamagnetic shift between the two configurations can be seen in the quantum well, since carriers are strongly confined in the growth direction. The diamagnetic shifts of quantum dots in both configurations are as small as that of a quantum well in the Voigt configuration, suggesting that strong 3D confinement is achieved in SAQDs. It is noticeable that the configuration dependence of diamagnetic shift is also seen in the PL from dots. This is consistent with the anisotropic shape of SAQD between the growth and lateral directions, as can be seen in the geometrical measurements (Fig. 1b).

3.3 Photoluminescence Excitation Spectrosopy

PLE spectroscopy is a powerful tool for investigating the relaxation mechanisms of carriers in quantum dots. While in typical PL the carriers are excited far above the conduction band edge, from whence they relax into dot states, in PLE carriers are created both in the dot and barrier states. For this purpose, variable wavelength is needed in the excitation light source. Here we use a Ti:sapphire laser. The excitation intensity is sufficiently low for carrier–carrier interaction effects to be small. Figure 5 shows ensemble PLE spectra at three different detection energies corresponding to the lower, center, and higher energy sides of the ensemble PL peak. One can expect that different emission energies correspond to dots with differing confinement. The arrows in the figure mark the excitation energy of 1.42 eV, at which we can see the absorption band-edge of the wetting layer (WL). Above the WL, states are dense since WL exhibit 2D-like structure.

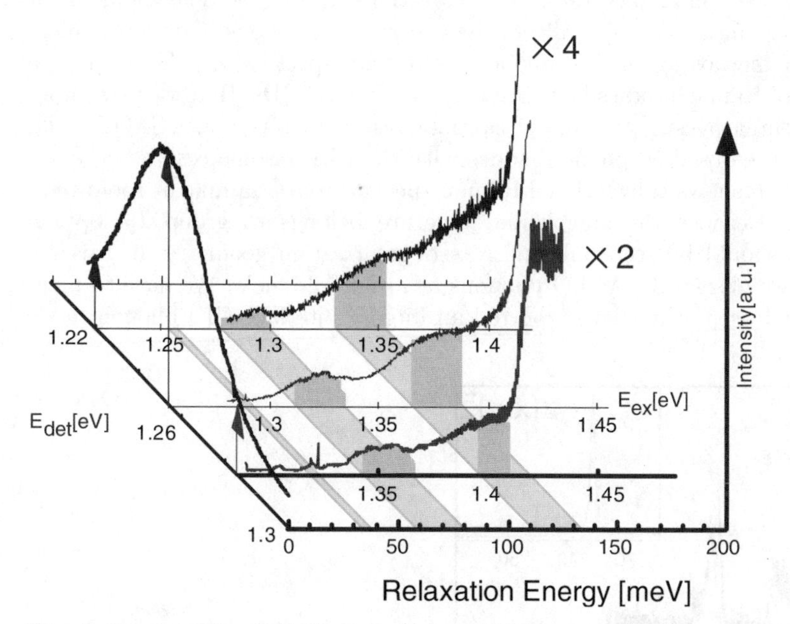

Fig. 5. 3D plot of far-field PL (*left side*) and PLE spectra (*main part*). PLE spectra are displayed at three different detection energies. The solid arrows in PL indicate the detection energies of corresponding PLE spectra. As can be seen, the relaxation energy remains independent of the excitation energy, indicating relaxation by phonons. The grey shadows in PLE show such relaxation energy-dependent resonances

In PLE spectroscopy, emission intensity (I_{em}) is related to excitation intensity (I_{ex}) by $I_{em} = P_{abs}P_{rel}P_{em}I_{ex}$, where P_{abs}, P_{rel}, and P_{em} are the probabilities of absorption, relaxation, and emission in the structure. P_{em} can be assumed to be constant in this measurement. In SAQDs, strong confinement produces sublevel separations much larger than acoustic phonon energies; P_{rel} is not constant due to the lack of phonon relaxation mechanisms between the sublevels. This indicates that absorbed carriers in states from which no relaxation channel exists cannot be a part of the PLE signal. Far-field PLE shows that the resonance energies do not depend on the detection energy but on the relaxation energy (E_{rel}), that is, the difference between the excitation and detection energies ($E_{ex} - E_{det}$). To summarize, one can understand that PLE resonances can occur at the energy of an integral number of phonons, in which efficient relaxation P_{rel} can occur.

3.4 Raman Spectroscopy

As mentioned in Sect. 2, optical phonons are important in considering carrier relaxation in SAQDs. Furthermore, the large lattice mismatch induces strain, which will change the lattice vibrations in SAQDs. The confinement and alloy composition also affects the energies of phonons. One direct measurement methos for optical phonons is Raman spectroscopy. Up to now, a number of Raman studies have been reported on SAQDs. Here we present an experimental investigation of optical phonons in InGaAs/GaAs SAQDs, and compare the observed phonon modes with the PLE resonances.

Figure 6 shows conventional Raman spectra from a sample at room temperature. Because the zincblende structure belongs to group T_d, optical phonons should be seen only in crossed polarization geometry. In this geometry, we observe GaAs LO phonons at around 36 meV. We have to mention that Pusep et al. have reported an intense interface (IF) phonon mode

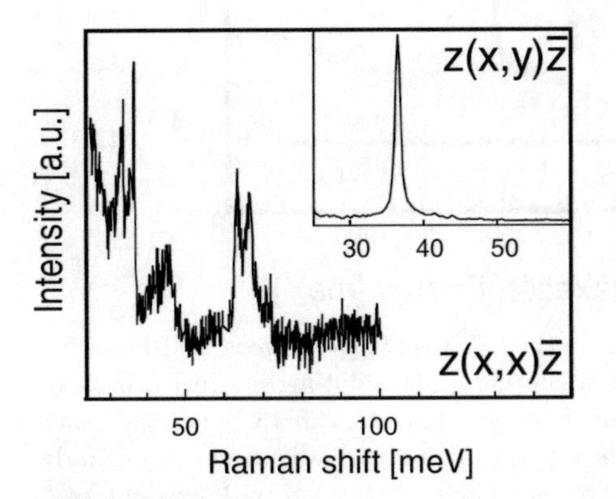

Fig. 6. Backscattered Raman spectra from [001] surface at room temperature. Excitation energy is 2.54 eV

in InAs/GaAs SAQDs [13]. Because the IF mode can exist at the discrete boundary, the out-diffusion of In atoms to the cap layer in our post-annealed sample can reduce the frequency gap between InAs and GaAs [14]. Another possibility is the difference of geometrical structure of SAQDs. While we can identify no phonon structures other than GaAs LO phonons in the crossed configuration, other phonon structures can be seen in the main part of Fig. 6, where the excitation and detection polarizations are the same; in this configuration we still expect localized phonon structures to be seen. The observed Raman structures are at 45 meV and 65 meV, and may be related to the PLE structures discussed in the previous section. We will reconsider these phonon modes in Sect. 4.2.

When we tune the laser energy to an electronic transition associated with the structure of interest, we can obtain enhanced Raman signals due to so-called resonant Raman scattering. Because the enhancement can reach several orders of magnitude, we can obtain the Raman signal from a small-volume structure such as a single epilayer. In addition, resonant Raman scattering provides information on electron–phonon interactions because enhancement strongly depends on the resonance conditions. Figure 7 shows Raman spectra excited in the vicinity of the WL absorption band edge. In the upper two

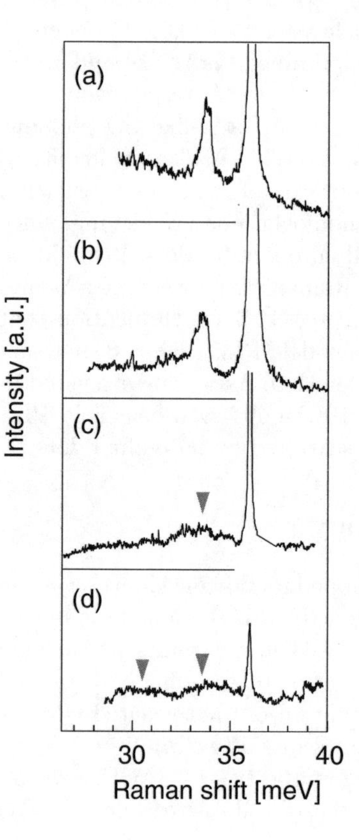

Fig. 7. Raman spectra under near-resonance conditions at $T = 8$ K. Crossed polarization configuration is employed. (**a**) $E_{ex} = 1.476$ eV, (**b**) 1.425 eV, (**c**) 1.333 eV, (**d**) 1.292 eV. The arrows indicate broad Raman peaks at ~ 33 meV and ~ 30 meV, respectively

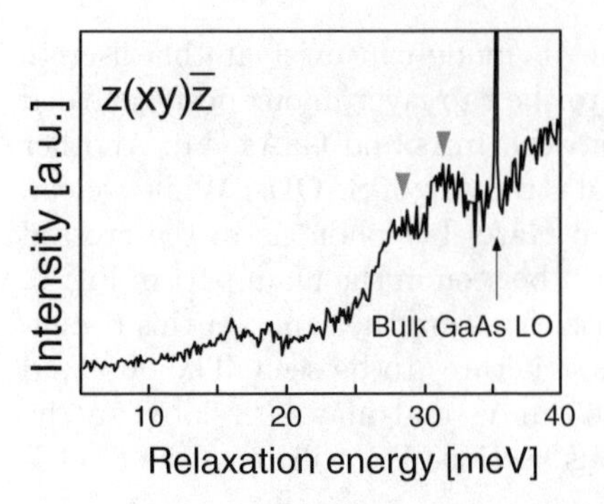

Fig. 8. A typical far-field polarized PLE spectrum around "1LO" region. Arrows are attributed to the GaAs-like (33 meV) and InAs-like (30 meV) phonon modes, respectively. Note that unknown weak structures can be seen below 25 meV

spectra, in addition to the GaAs LO phonon, the transverse optical (TO) phonon mode can be seen around 34 meV because of the breakdown of the selection rules in the resonance condition. With decreasing laser energy, an additional broad but distinct peak appears at 33 meV. In Fig. 7d, where carriers are excited in the dot states, we can see a second broad peak at $\sim$ 30 meV. For strain-induced $In_xGa_{1-x}As$ layers ($x > 0.5$), Groenen et al. have performed Raman spectroscopy and obtained GaAs-like and InAs-like LO phonon energies of $<$ 33 meV and $>$ 29 meV, respectively [15]. Note that they also observed that the linewidth of GaAs-like LO phonons is significantly greater than that for bulk GaAs. This is also analogous to our observed phonon behavior. Moreover, in our case of the SAQD structure, dot-related optical phonons might exhibit inhomogeneous broadening, since these will be strongly scattered in the individual quantum dots. It might be reasonable to attribute these observed broad Raman structures to GaAs and InAs related optical phonons in SAQD structures. It is worth mention that these phonon peaks clearly can be seen in the far-field PLE. Figure 8 shows an expanded view of far-field PLE in the region of 1LO. As will be discussed in Sect. 4.2, this conclusion is also consistent with the observed near-field PLE spectra, in which corresponding features are scattered by individual dots.

4 Near-Field Optical Spectroscopy

Near-field optical spectroscopy has been developed for advanced investigation of nano-structures with high spatial resolution. In this section, we describe and discuss several characteristic phenomena based on the results of near-field experiments, including the dependence on excitation power, temperature, and external magnetic field. Considering the relaxation of photoexcited carriers in SAQDs, we carry out near-field PLE spectroscopy as well.

Before discussing about near-field optical properties of SAQDs, we introduce our low-temperature scanning near-field optical microscope (SNOM)

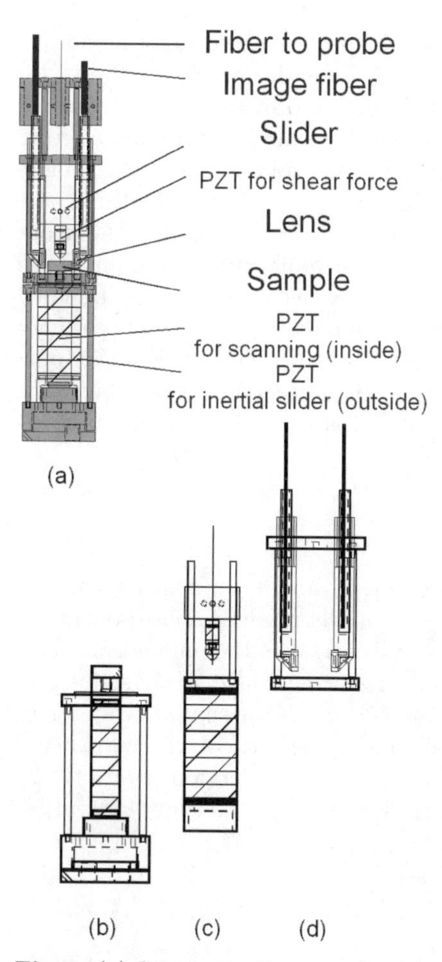

Fig. 9. (**a**) Schematic diagram of our low-temperature SNOM head. Three piezo-electric transducers (PZTs) are used: one dithers the fiber probe for shear-force, another is a slip-stick motion for coarse approach of the probe to the sample, and the third is for scanning the sample. Shading indicates different materials. The SNOM head consists of three parts of (**b–d**)

[16]. The whole scanning setup is installed in a helium gas-flow cryostat within a 2–1/4 inch diameter sample chamber, which is mounted on a vibrationally and magnetically isolated optical table. For magneto-optical spectroscopy, we use an SNOM incorporating a superconducting magnet, giving fields up to 10 T. The design of our custom-built SNOM head is shown in Fig. 9. In order to minimize the change of tip position in high magnetic field, all parts were made of non-magnetic materials. The SNOM apparatus consists of three main parts. One is the scanning head (Fig. 9b), which allows the sample to be scanned in lateral directions. The second is the mount of the fiber probe,

consisting of an inertial slider that enables the probe to approach the sample, and a dithering piezoelectric transducer (PZT) for shear-force feedback [18] (Fig. 9c). Piezoelectrically driven slip-stick motion provides for coarse movement of the tip [17]. In this motion, we need no mechanical contact from outside the cryostat. Therefore the position of the tip relative to the sample position can be kept unchanged even in high magnetic fields. The third part is the optical access to the sample (Fig. 9d). We use a cryostat without optical windows, so for imaging both the tip and sample, and to illuminate the 1.3-μm laser diode light for shear-force measurements, we employ bundled fibers [19]. In the analysis of semiconductor nano-structures, to obtain high spatial resolution, carrier diffusion effects should be avoided. For this purpose the illumination–collection mode (I-C mode) SNOM, where both the excitation and collected luminescence light pass through the same fiber, is employed [20,21].

4.1 Ground-State Emission

Radiative recombination in the excitonic ground state reveals various optical properties in SAQDs. Figure 10a shows a spatially resolved non-resonant PL spectrum at liquid helium temperatures. In contrast to the inhomogeneously broadened PL spectrum in the far field, PL in the near field shows spiky lines in the corresponding energy region. Figure 10b shows the spatial evolution of such near-field PL spectra, where the horizontal and vertical axes correspond to the detection energy and probe position in one dimension, respectively, and white color indicates intense light. In these data distinct luminescent re-

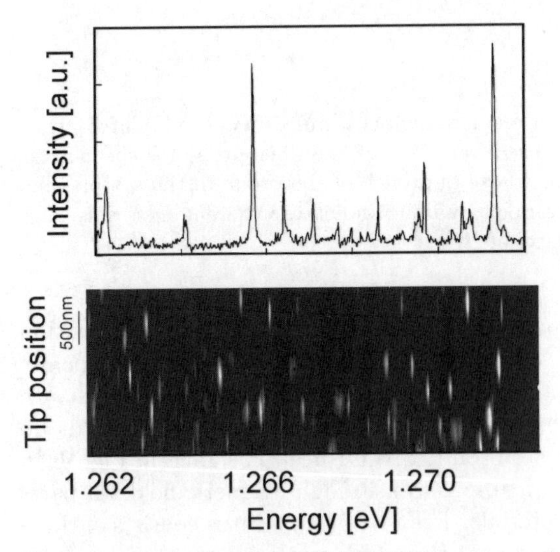

Fig. 10. (**a**) A typical near-field PL spectrum excited by a HeNe laser. (**b**) Spatial evolution of the near-field spectra in one dimension

gions from individual dots can be resolved, and thus the sharp lines observed in Fig. 10a can be identified as the emission from single dots. As has been reported by many other groups, each line observed in this sample has a narrow linewidth, about 50 – 100 μeV, which is slightly greater than the monochromator resolution of around 50 μeV. Far-field time-resolved PL measurements at 10 K suggest the carrier lifetime to be longer than 600 ps, corresponding to a FWHM lower limit of several μeV. Therefore, the observed linewidth involves the phase relaxation of carriers in ground-state recombination.

In order to evaluate the observed homogeneous linewidth, we begin by considering scattering effects between carriers. In Fig. 11, we plot the excitation power dependence of a typical PL peak under non-resonant conditions. The luminescence shows the same peak energy at different excitation power. This indicates that thermal heating by the illumination light is absent be-

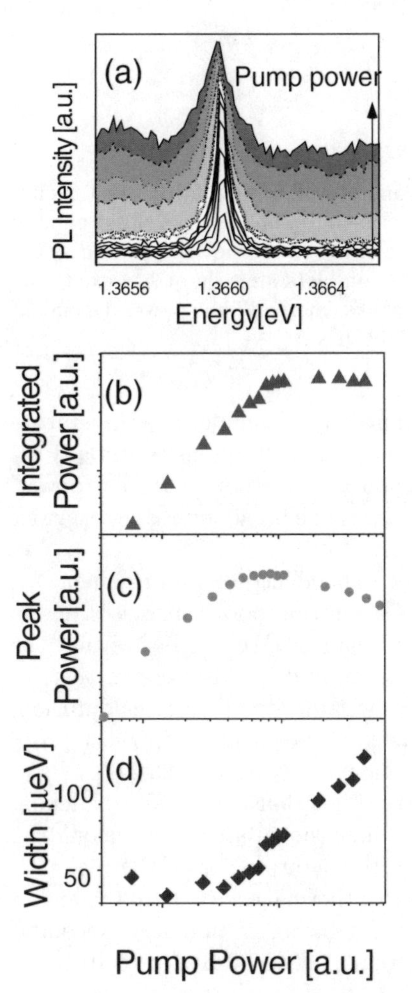

Fig. 11. (**a**) A typical single dot PL spectra excited by a HeNe laser at various pump powers. (**b**) The plots show integrated power, peak power, and linewidth of the PL peak in (**a**) as a function of the excitation power

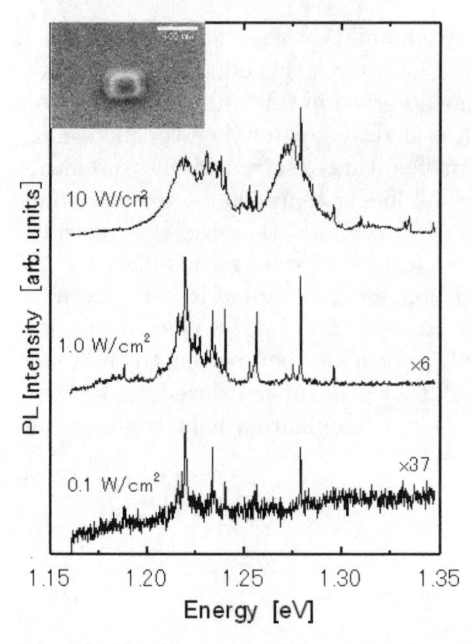

Fig. 12. Non-resonant μ-PL spectra from island-etched SAQDs at different excitation powers [22]. The inset is the SEM image of the sample, so a limited number of dots are excited. The growth conditions are the same as the sample used in the near-field experiments. Several distinct peaks from dots are seen at low excitation power. At higher excitation powers, multiple peaks emerge. These power dependent multiple peaks are a characteristic feature of SAQDs

cause the emission energy is drastically changed by changing the temperature as discussed later. As the pump power is increased, PL intensity initially increases linearly, and then reaches saturation at a certain excitation power. This saturation suggests band-filling and causes line broadening due to carrier scattering.

Figure 12 shows the excitation power dependence of non-resonant PL spectra from an island-etched sample. As excitation power increases, many peaks appear and the spectrum gradually broadens. We emphasize that the number of excited quantum dots is strictly limited in this sample; there is no inhomogeneous broadening of the PL spectrum due to carrier diffusion. Therefore both of multiple peaks and broadened spectrum are intrinsic properties of SAQDs and can be explained by many-body effects [23,24].

Because of full quantization in all three dimensions, thermal broadening of recombination states as seen in higher-dimensional structures should be absent in quantum dots. However, elevated temperature causes thermal activation of electron–phonon interactions and/or thermal excitaion to the higher excited states, and results in line broadening during ground-state recombination [25]. Figure 13 shows PL spectra from 3.7 K to 90 K. Up to 10 K, no

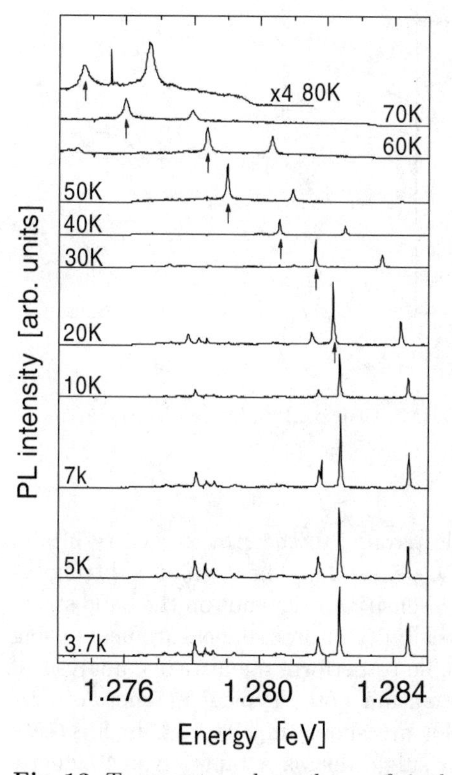

Fig. 13. Temperature dependence of single-dot PL spectra. The spectra are shifted vertically for clarity. The peak of interest is indicated by arrows

change is observable either in emission energy or linewidth. Above 20 K, PL peaks show a redshift and line broadening almost simultaneously. The quantized energy of dots depends on the energy band gaps. Therefore the observed lower energy shift in single-dot peaks can be attributed to the temperature dependence of the band gap. In fact, the Varshini relationship using bulk parameters can fit the data well, while the effect of carrier–phonon coupling in dots should be taken into account for detailed analysis. The line broadening is characterized both by population decay and dephasing in ground-state recombination. In Fig. 14, we plot the PL linewidths as a function of temperature. Because in SAQDs confinement on length scales down to 10 nm in all three directions produces sublevel separations of about 50 meV, we can rule out thermally excited carriers at low temperature. Takagahara has demonstrated the presence of pure dephasing in a 0D system, and that the electron–acoustic phonon interactions dominantly contribute to the temperature dependence of pure dephasing [27]. Therefore the observed line broadening may reflect the temperature dependence of the pure dephasing rate of ground-state recombination in SAQDs [26].

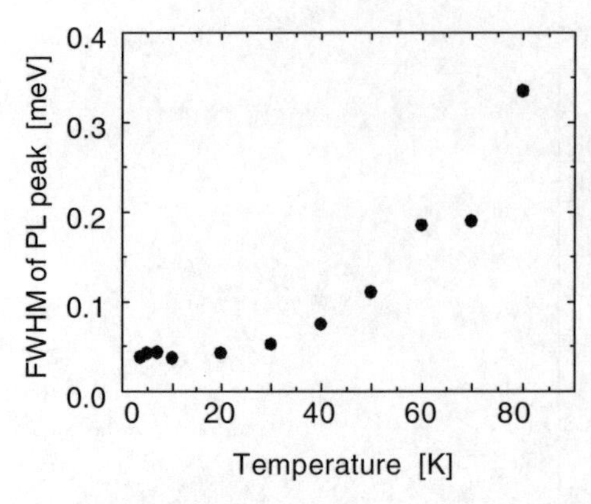

Fig. 14. Temperature dependence of FWHM of PL peak indicated by the arrow in Fig. 13

The application of a magnetic field parallel to the growth axis results in Zeeman spin splitting of individual PL peaks [28]. The Zeeman splitting reflects the spin magnetic moment, which sensitively depends on the band structure and carrier wave function. The feasibility of precise measurements using single-dot spectroscopy makes it possible to perform quantitative analysis of carrier wave functions confined to quantum dots. Typical magneto-optical luminescence spectra from a single dot are shown in Fig. 15a. In Fig. 15b, we plot the splitting energies from a single dot as a function of magnetic field. The splitting of the peaks changes linearly with magnetic field, which clearly demonstrates that the splitting results from the Zeeman effect. The

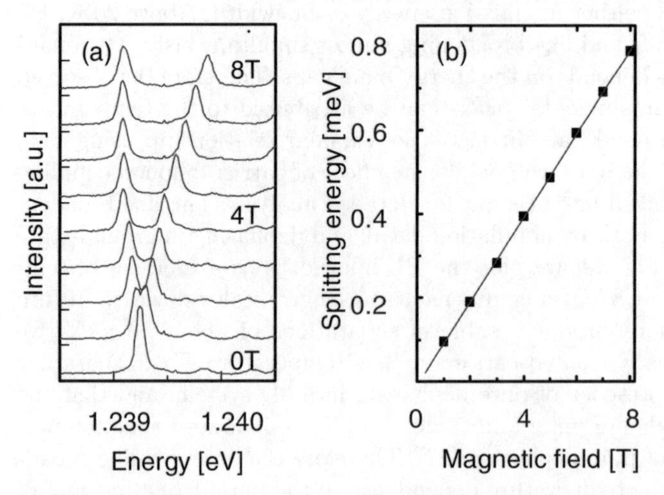

Fig. 15. (**a**) Typical unpolarized photoluminescence spectra from a single dot at various magnetic fields. (**b**) Magnetic field dependence of the splitting energy

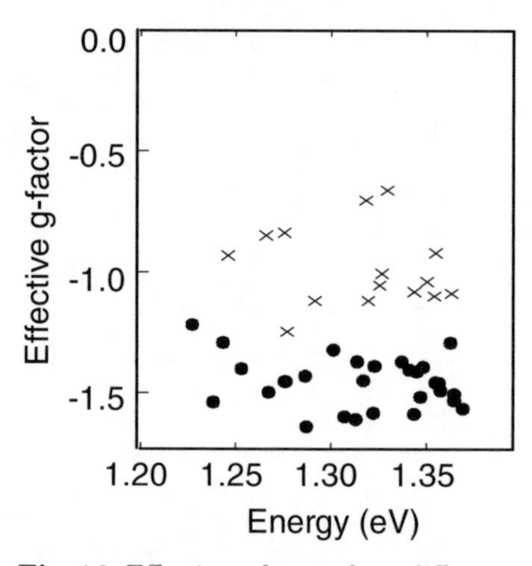

Fig. 16. Effective g-factors from different single-dot PL as a function of the emission peak energy at 0 T. Cross denote results from an $Al_{0.36}GaAs_{0.64}$ capped SAQD sample

magnitude of the energy splitting ΔE can be expressed as $g^* \mu_B B$, where g^* is the effective g-factor and μ_B is the Bohr magneton. The linear coefficient is 85 μeV/T, which corresponds to a g^* of -1.47. The minus sign can be determined by polarization spectroscopy, as discussed below. We comment on the observed g^* of ~ -1.5. For additional information about g^* in the present sample, we investigate the distribution of g^* for ground-state emission. Figure 16 shows the magnitude of g^* for individual dots with various emission energies in the same sample, plotted as a function of the PL peak energy in zero magnetic field. The distribution of the emission energies ranges over 100 meV. In order to compare systems with different confinement, the Zeeman splittings for dots capped by $Al_{0.36}GaAs_{0.64}$ are also plotted in the figure. There is no clear energy dependence of g^* in the same sample, while different confinement systems exhibit different averaged values. For SAQDs, strong variation of g^* might be expected in each dot because of the large inhomogeneity of the size, strain, and composition. However, the observed non-scattered g^* indicate that inhomogeneity of SAQDs does not affect the spin momentum strongly. Similar non-scattered g^* has also been observed in the single dot PL with different dopants [29].

Near-field imaging of individual quantum dots in the magnetic field allows direct observations of spin states in ground-state emission. Recombinations of different spin states generate PL light with different circular polarizations. Therefore spectroscopy with polarization discrimination provides information about the spin states. Such polarized PL spectra can be successfully obtained by using linearly polarized excitation and a quarter-wave plate for collection. Figure 17 shows NSOM images of the two different circular polarizations at 6 T. As indicated in the figure, one of the two split peaks in the same position

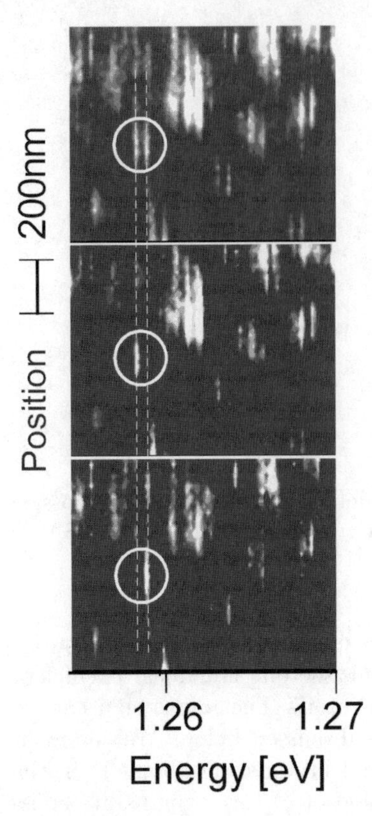

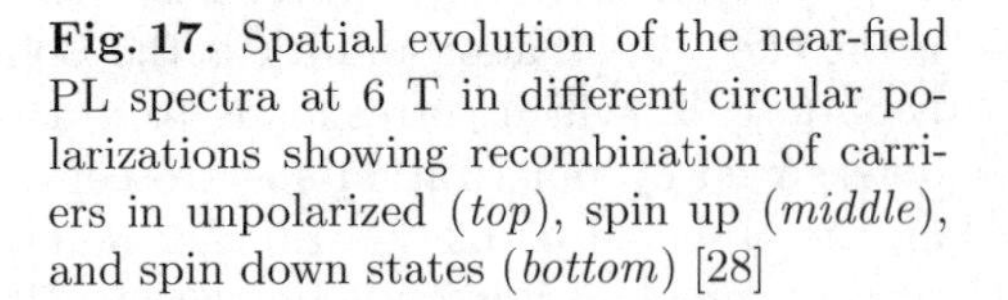

Fig. 17. Spatial evolution of the near-field PL spectra at 6 T in different circular polarizations showing recombination of carriers in unpolarized (*top*), spin up (*middle*), and spin down states (*bottom*) [28]

is suppressed in each polarization, thus showing the existence of spin states in the single SAQDs. The results indicate that the higher peaks originate from spin down carriers, and the lower from spin up. Therefore, g^* is negative. There still remains a small PL peak corresponding to the suppressed Zeeman component because the polarization selectivity of the experimental setup is not complete.

The magnetic field dependence also indicates that the PL peak exhibits a diamagnetic blue shift: this is shown by the average value of the energies of the two split peaks. In the low magnetic field regime, the energy shift of QD energy levels is given by $\Delta E = \beta B^2$, with diamagnetic coefficient $\beta = e^2 \langle x^2_{e(h)} \rangle / 4m^*{}_e(h)$, where m^* is the effective mass of electron and hole, and $\langle x^2 \rangle$ gives the spatial extent of the carrier wave function in the lateral direction. By using a reduced mass $m^* = 0.08 m_0$, evaluated from the magnetic field dependence of PL of a InGaAs/GaAs quantum well sample, the observed diamagnetic shift of $\sim 4\ \mu\mathrm{eV/T}^2$ corresponds to a carrier wave function of lateral extent ~ 5 nm. This result indicates that carriers are strongly confined in the lateral direction in excitonic ground states of the SAQDs.

4.2 Interaction with Phonons

In various optical properties of SAQDs, one can observe contributions due to fully quantized states with large energy separation. For example, near-field PLE spectra show an extremely large no-signal region above the ground-state emission, as shown in Fig. 18. On the other hand, the PLE shows various unique features between the zero-absorption and the WL absorption band edge (see Fig. 18a). The sharp resonance lines observed in each near-field PLE spectra occur within the same relaxation energy region, indicating independence of the detection energy, as observed in ensemble PLE spectra. Considering the relaxation energy dependence, PLE resonant peaks are predominantly due to resonant phonon Raman scattering. Here we will consider electron–phonon interactions on the basis of observed near-field PLE resonances. By analogy with the far-field PLE, we can classify PLE structures into two parts, namely the 1LO and 2LO regions. First, we concentrate our interest on the 1LO region, located below the relaxation energy of 50 meV.

Because zero absorption in PLE ranges over 50 meV, which is greater than the energy of optical phonons, resonant Raman features of optical phonons can be measured. Figure 18b shows a expanded view of the 1LO region, in which several weak and narrow resonant features are clearly seen. With extremely narrow linewidths observed in single SAQD emissions, resonant Raman spectra provide sensitive measurements of local environments around

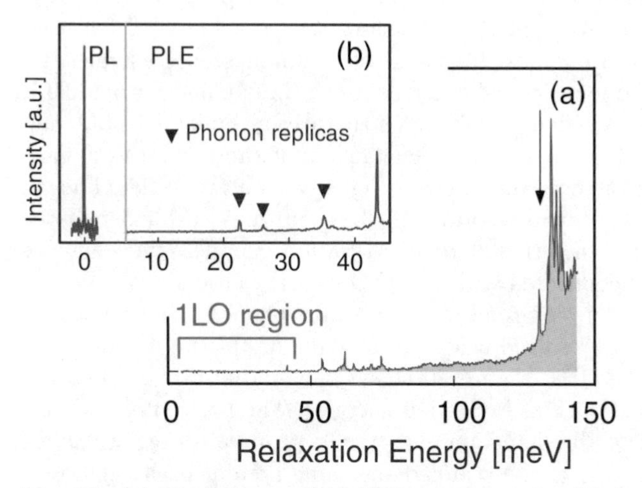

Fig. 18. (**a**) Typical near-field PLE spectrum of a PL feature as shown in (**b**). The detection energy $E_{det} = 1.282$ eV. The arrow indicates the WL absorption band edge around $E_{ex} = 1.42$ eV. (**b**) Expanded view of "1LO" region. Because Lorentzian fitting of the PL feature yields a linewidth as narrow as 60 μeV, the resolution of PLE is determined by the scanned steps of the laser energy > 150 μeV. The shading in (**a**) indicates the difference from the background

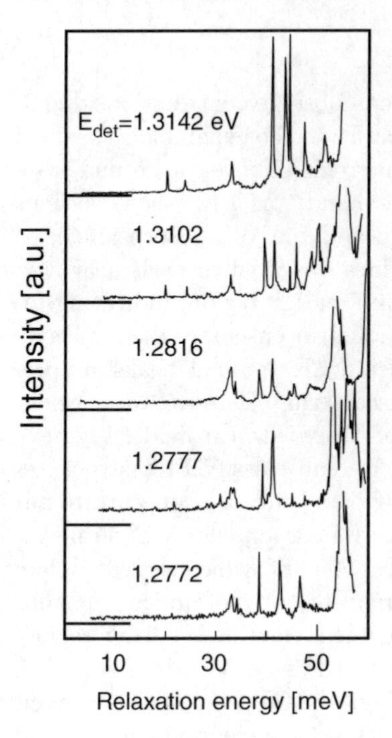

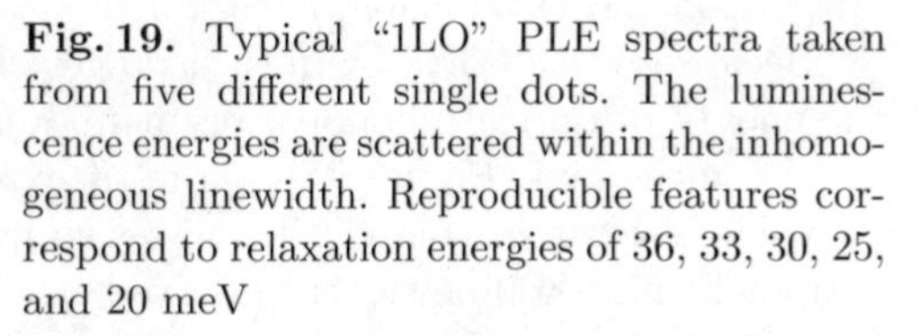
Fig. 19. Typical "1LO" PLE spectra taken from five different single dots. The luminescence energies are scattered within the inhomogeneous linewidth. Reproducible features correspond to relaxation energies of 36, 33, 30, 25, and 20 meV

individual SAQDs. Figure 19 shows five typical PLE spectra in the 1LO region. In these spectra some of the dots show reproducible PLE features. Because these features are almost independent of luminescence energies, we suppose that they are due to optical phonons. Both Raman and far-field PLE spectroscopy allow us to assign the phonon structures as bulk GaAs LO, and dot-related GaAs-like LO for relaxation energies of 36, and 33 meV, respectively, while other resonances on the lower energy side are still unknown. It is worth noting that resonances around $E_{rel} = 40$ meV also exist at the same relaxation energy. In terms of their relaxation energies, they may be attributed to the 2LO connected with the 1LO existing around 20 meV.

In order to obtain more information on electron–phonon interactions, we next carry out spectrally resolved near-field imaging of optical phonons coupled with individual SAQDs. Figure 20 shows the spatial evolution of the near-field PL spectra at various excitation energies. The horizontal and vertical axes correspond to the detection energy and probe position in one dimension, respectively. The probe scanned the same area at each excitation energy. Under non-resonant conditions, regions giving single-dot luminescence emit at energies scattered throughout the whole range of detection energies (Fig. 20a). The relative intensities of the emission lines are independent of excitation energy until the laser is tuned close to the 2LO energy, at which point they become very sensitive to resonance (Fig. 20b). In Fig. 19 a reproducible peak can be seen at $E_{rel} = 36$ meV. This 1LO resonance makes

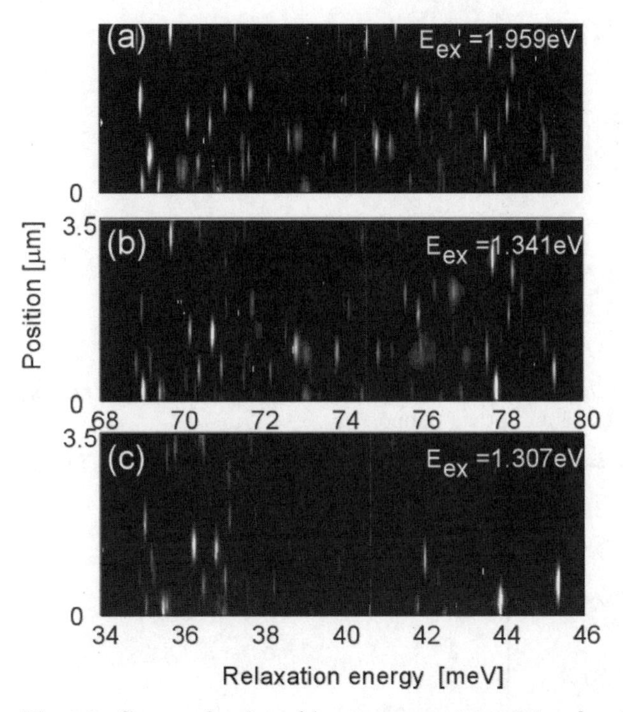

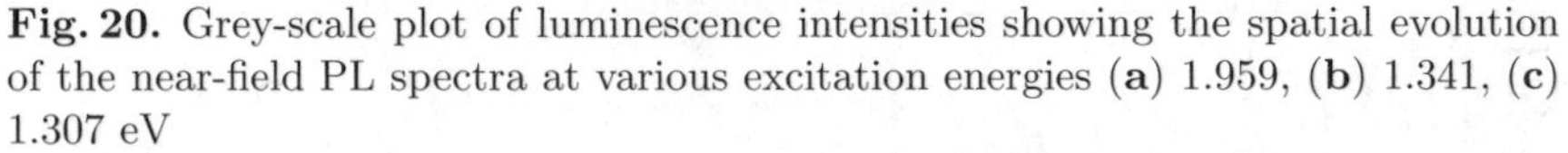
Fig. 20. Grey-scale plot of luminescence intensities showing the spatial evolution of the near-field PL spectra at various excitation energies (**a**) 1.959, (**b**) 1.341, (**c**) 1.307 eV

ground-state emission visible in Fig. 20c. In this figure, almost all PL features show bright resonances, thus suggesting that the 1LO feature observed in PLE is due to a resonant Raman process in which GaAs-related phonons are emitted.

The results thus suggest strong interaction between electrons and phonons in SAQDs. Note that a bulk GaAs resonance is also observed in InAs SAQD samples, where the GaAs phonon exists in the cap layer. On length scales down to several tens of nm, the exciton wave function penetrates the cap layer and enhances the interaction with the phonon mode in the barrier.

Except for the peak of GaAs-related LO phonons, some of the dots show PLE features around 30 meV. In contrast to the reproducible GaAs related phonons, the exact energies of these 1LO features are different in each near-field PLE spectrum. This variation in the peak positions and intensity of the phonon lines in each spectrum can be attributed to inhomogeneity in the distribution of SAQDs. As discussed in Sect. 3.4, we identified broad Raman structures as an InAs-related phonon, which is scattered due to the distribution of both strain and compositions. Therefore 1LO resonant peaks reflect local environments around individual SAQDs. In Fig. 21, we summarize a schematic illustration of the observed resonant Raman process based

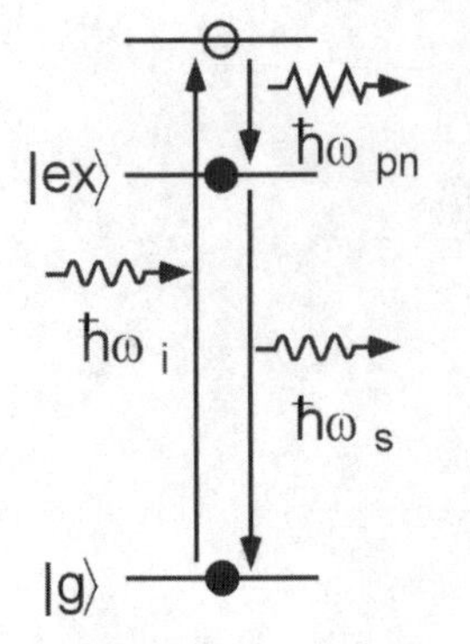

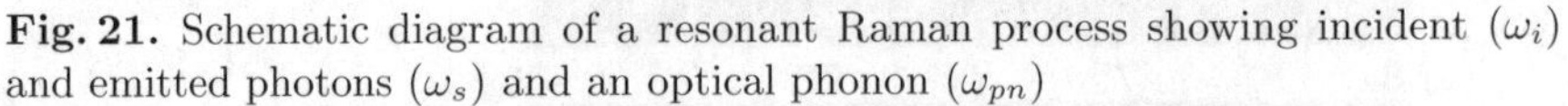

Fig. 21. Schematic diagram of a resonant Raman process showing incident (ω_i) and emitted photons (ω_s) and an optical phonon (ω_{pn})

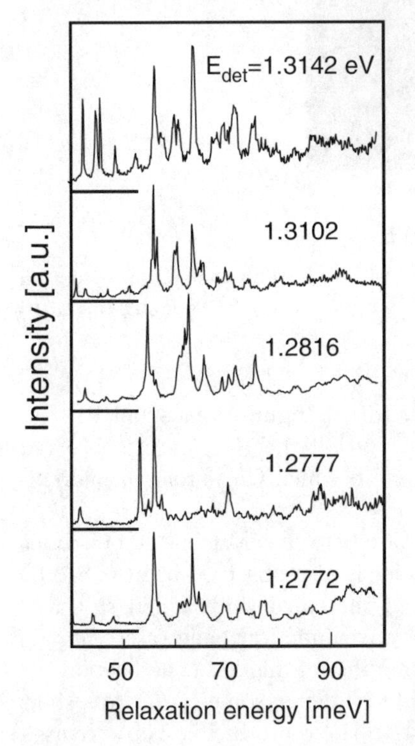

Fig. 22. Typical "2LO" PLE spectra taken from five different single dots

on phonon emission. When the relaxation energy matches the energy of an available phonon, phonon generation satisfies the resonance condition in the ground state emission and results in a resonant peak in the PLE spectrum.

In the so-called 2LO resonance region, the near-field PLE spectra show sharp resonance features. Figure 22 shows magnified PLE spectra around the 2LO region, in which a number of partly overlapping sharp resonance lines are found. The sharp resonance lines are observed within the same relaxation energy region in each near-field PLE spectra, showing an independence of detection energy, as observed for the 2LO resonance in the far-field PLE spectra. In contrast to the reproducible 1LO signal, features in the 2LO region are different in each near-field PLE spectrum: although the sharp resonance lines always occur within the relaxation energy range of the broad resonance observed in far-field PLE, their number and energies are different in each spectrum. It might be reasonable to attribute the observed Raman structures to localized phonons, since these will be enhanced and strongly scattered in individual SAQDs, due to the lack of long-range order of the lattice structure.

The observed multi-phonon related process is consistent with a reproducible 1LO GaAs phonon, and far-field PLE and Raman spectroscopy provide support for this conclusion. However, there is still a question why such intense 2LO peaks are observed in PLE. In fact, we observed virtually no signal under outgoing resonance conditions. We discuss here the origin of the enhanced resonance conditions observed in PLE.

In addition to the phonon resonance peaks, single-dot PLE spectra show an extensive continuum background which, starting around $E_{rel} = 50$ meV, gradually increases up to the WL absorption band edge. In Fig. 23, the plot of the PL intensity as a function of relaxation and detection energies indicates the absence of background emission at other wavelengths, and thus that the observed continuum is an intrinsic feature of single SAQD emission. The existence of the continuum states may be connected with the WL. It is plausible that the potential of the WL outside the dots is different from that of the WL within dots. In other words, the potentials of dots and WL are affected by the strain in each other. Even "stressor dots" can be fabricated in 2D

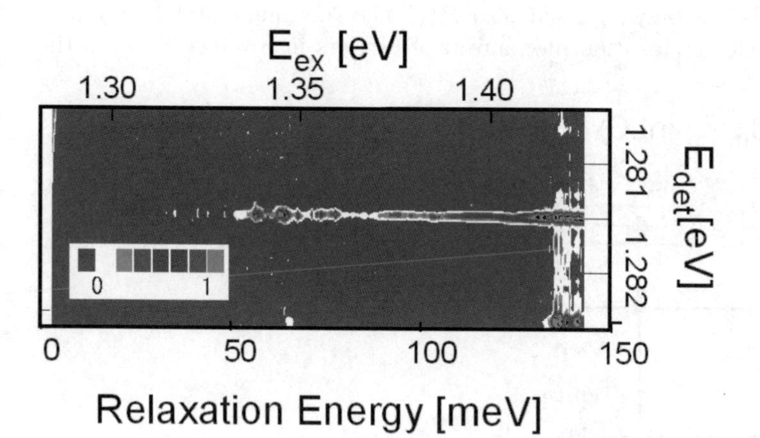

Fig. 23. Grey-scale contour plot of near-field PL intensities as a function of detection and excitation energy. Relaxation energy corresponding to the PL peak is also displayed

quantum wells by the strain induced by SAQDs [31]. These results suggest that the potential of the dots connected with WL might be strongly modified by strain. Based on these assumptions, SAQDs may have a crossover between 2D and 0D character [30]. It is important to note that such a continuum has also been observed in many other reports on single SAQDs [32–34].

The existence of continuum states can explain the intense 2LO phonons under incoming resonance conditions. In PLE spectra we observe that the intensity of 2LO peaks is much stronger than that of 1LO, suggesting that the phonon lines observed around 65 meV involve enhanced scattering cross sections. We suggest that the continuum states provide enhanced scattering cross sections for phonons. Further investigations are needed to clarify the resonance conditions. However, we stress that the observed 2LO lines involve strongly enhanced scattering cross sections due to strong electron–phonon interactions.

4.3 Carrier Relaxation

In the previous section, we refer to the contribution of continuum states to the PLE signal. In addition, near-field PLE spectra also show a number of sharp resonances, which can be understood as resonant Raman processes associated with localized phonons due to the loss of long-range order of the lattice in SAQDs. Based on these measurements, we suggest following three processes as possible relaxation mechanisms of excited carriers: (a) relaxation through the continuum via acoustic phonons; (b) multiple-phonon relaxation; (c) interlevel relaxation (see Fig. 24). It is worth to noting that the 2LO region should involve higher excited states. However, the excited states are obscured by the continuum states and resonant phonon structures, because the PLE signal reflects not only P_{abs} but also P_{rel}. The continuum states may provide the efficient relaxation mechanism of carriers in SAQDs. Because the

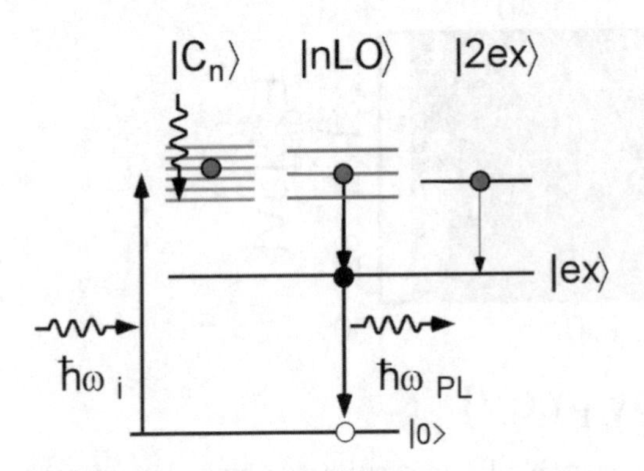

Fig. 24. Three possible carrier relaxation mechanisms in SAQDs: (*left*) relaxation through the continuum ($|C_n\rangle$) via acoustic phonons; (*center*) direct transition via multiple-phonon ($|\mathrm{nLO}\rangle$) emission; (*right*) interlevel relaxation

continuum states connect with the WL states, the carriers excited in the non-resonant case can relax efficiently by way of these states. As discussed in the previous section, the observed intense 2LO lines reveal strong electron–phonon interactions. These interactions should also lead to efficient relaxation involving LO phonons under non-resonant conditions. Because the continuum states extend right down to the 2LO part of the spectrum, these resonances can provide efficient relaxation channels for ground state emission. Our results thus indicate that intradot relaxation through the continuum states is more important than interlevel relaxation through discrete confined states in SAQDs. We believe that the existence of the observed continuum states provides a plausible interpretation for why efficient relaxation is achieved in SAQDs. Moreover, based on this mechanism, efficient relaxation of carriers can be achieved in SAQDs even in the presence of the phonon bottleneck.

We comment on the contribution of the higher excited states of SAQDs. Ensemble PL spectra suggest that there may exist two other excited states whose energies are separated by 50 and 100 meV from the ground state emission (Fig. 2). Therefore observed PLE features around 2LO region involve higher excited states. Because the Zeeman splitting of the higher excited states shows different behavior from that of the ground state due to the additional orbital angular momentum [35], we can assign the higher excited states. Figure 25a–b shows typical PLE spectra with and without a magnetic field, respectively. PLE for different split branches of a PL peak are shown in Fig. 25a. In the PL peak the ground state, Zeeman splitting is estimated to be 1 meV at 8 T. Figure 25c shows for each PLE resonance the difference between the energy splitting observed in PLE and that observed in PL. For most PLE resonances, the relaxation energy is independent of the magnetic field. This result agrees well with the explanation of the PLE resonances as resonant Raman scattering from phonons, because the phonon only reflects the ground-state energy, so such peaks should show the same magnetic field dependence as the PL peak. Nevertheless, some PLE peaks show small changes in their splittings, suggesting Zeeman behavior of the higher excited states of dots. A similar small change of higher excited spin splittings has been observed for different 0D systems [36]. In Fig. 25, the peak with different Zeeman behavior exists around $E_{rel} = 43$ meV. This relaxation energy is consistent with the higher excited states estimated from ensemble PL.

4.4 Dephasing of Excited Carrier

Since the DOS structures in SAQDs have large energy level spacing in which acoustic phonon-mediated scattering should be suppressed, it is predicted that carriers in excited states may exhibit long decoherence times, which is advantageous for implementing coherent control of the quantum state [37]. On the other hand, optical spectroscopy suggests that the interaction with LO phonons plays a crucial role in the inter-level relaxation, as discussed in the previous section.

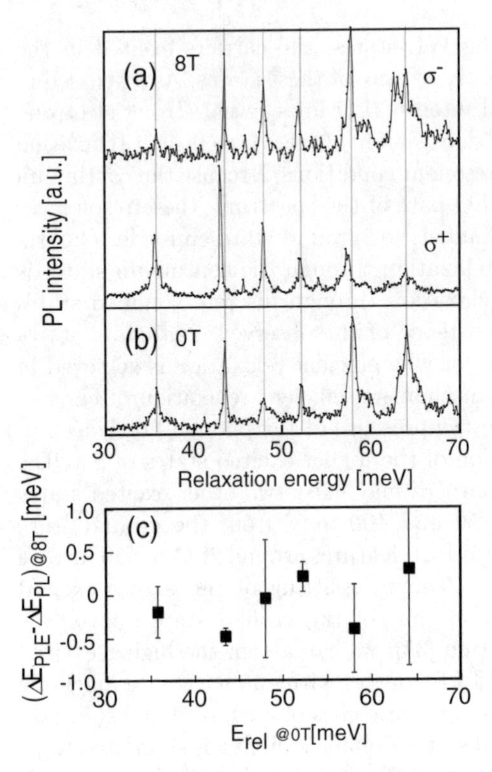

Fig. 25. PLE spectra from a single dot at (**a**) 8 T and (**b**) 0 T. Six typical resonance peaks are plotted as a function of relaxation energy. The grey lines indicate the peak position at 0 T for clarity. (**c**) Difference of splitting energies from PL splitting at each PLE resonance observed in (**b**)

In order to investigate the phase relaxation of excited carriers, we demonstrate with near-field coherent excitation spectroscopy [39]. For this experiment, we employed wave packet interferometry, which has been developed for the coherent optical control of monolayer 0D islands in quantum wells by N. Bonadeo et al. [38]. Figure 26 shows our experimental setup based on SNOM for coherent excitation spectroscopy. A mode-locked Ti:sapphire laser is used as a light source. In order to reduce the group velocity dispersion in the optical fiber, output pulses go through a grating pair, which reshapes the pulsewidth. This also provides optimum spectral linewidth ($\sim$ 3 meV), which is sufficiently narrow to excite only one resonance (see Fig. 27).

Two phase-locked pulses with a time delay between them are created by a Michelson interferometer, and they excite the carriers. When the laser energy is tuned to one of the resonances, the integrated PL intensity shows oscillations as a function of the delay time, as shown in Fig. 28a. During phase correlation of the scattered carriers, oscillations can be observed in the PL intensity. Therefore, in this measurement we can directly observe the dephasing

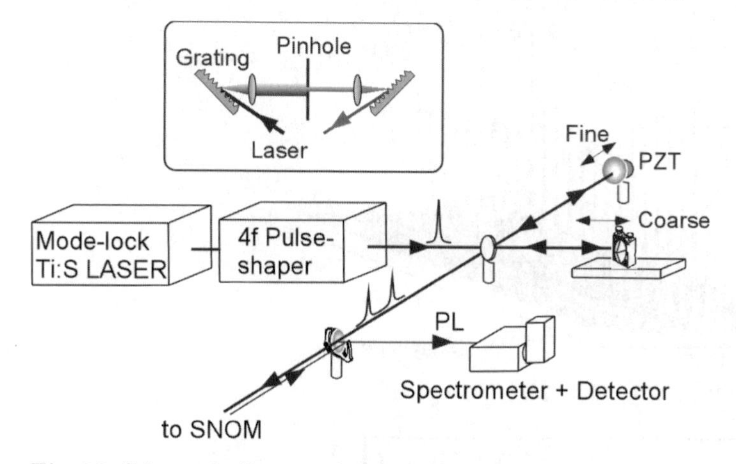

Fig. 26. Schematic diagram of the experimental setup for near-field coherent excitation spectroscopy. The square indicates the 4f-pulse shaper for optimum excitation

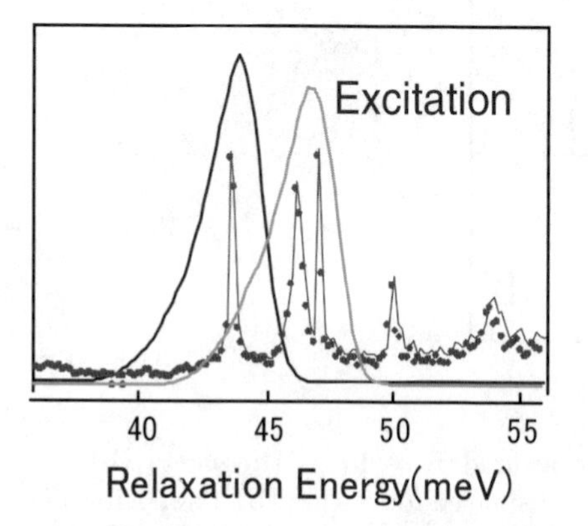

Fig. 27. PLE from a single dot around $E_{rel} = 40$ meV on an expanded scale. The solid line shows the laser spectrum, whose linewidth is sufficiently narrow to not excite other peaks. The grey line also shows the laser spectrum, but for coherent superposition of carrier wavefunctions [39]

time of the relaxation process. The time resolution of this system, including fiber, is estimated to be around 1 ps from the autocorrelation function of the laser pulse. Note that we monitor the intensities of luminescence at energies far from the excitation energy, so that the interference signal is independent of the laser light, even in the copropagating geometry used in this experiment (see Fig. 28a).

In Fig. 28b are plotted the time-integrated PL intensities as a function of delay time. The large circles indicate the oscillation amplitude as determined by fitting the data with a sinusoidal function. With increasing time delay, the amplitude of the oscillation decreases according to the decay of the

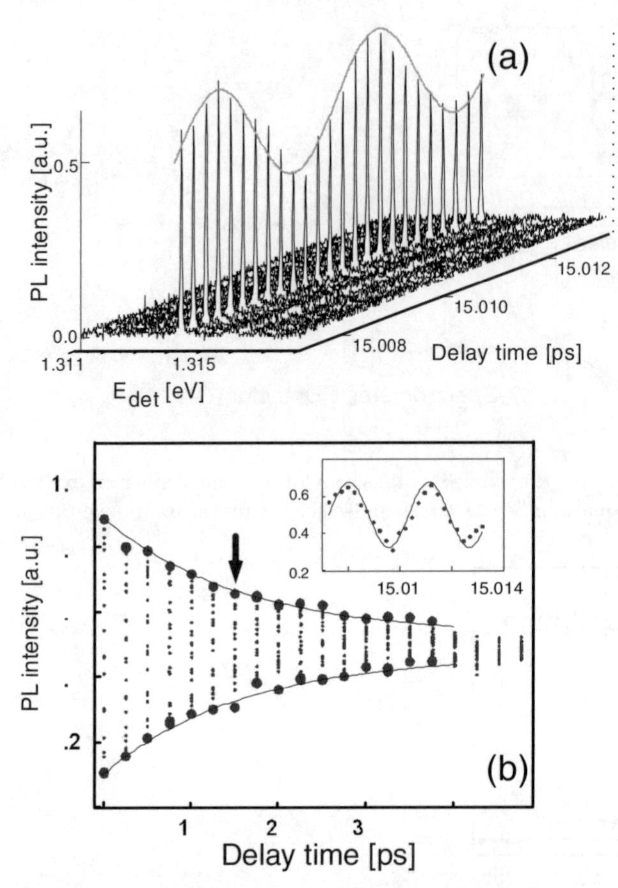

Fig. 28. (**a**) Time-integrated PL spectra around $\tau = 15$ ps (*arrow* in (**b**)). A sinusoidal fit through the data points is also displayed for clarity. (**b**) The oscillation amplitude of the time-integrated PL as a function of pulse delay. Inset: Interference oscillations around 15 ps on an expanded scale with sinusoidal fits through the data points [39]

coherence of excited carriers. An exponential fit yields the dephasing time of the resonance to be about 15 ps. This value is almost equal to the inverse linewidth obtained by Lorentzian fitting of the PLE resonance.

Let us consider the dephasing time of 15 ps observed in this sample. In the case of inter-level relaxation between atom-like quantized states, we may expect a long dephasing time. That our measured dephasing time is significantly different implies that inter-level relaxation does not play a role in the PLE resonances. We consider the observed fast decoherence time in SAQDs to be consistent with the phonon-related relaxation discussed in the previous section.

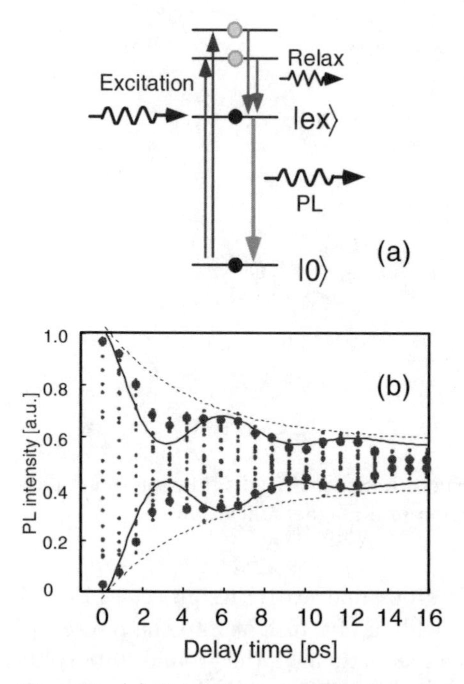

Fig. 29. (**a**) Schematic diagram for coherent superposition of carrier wave functions. By using two closely spaced resonances, coherent superposition of carrier wavefunctions with different energies can be generated. In this measurement, we tuned the laser to an energy at which double peaks can be excited simultaneously (grey line in Fig. 27). (**b**) The quantum interference between two resonances, which may be attributed to virtual states coupled to the localized phonons [39]

Coherent superposition of carrier wavefunctions with different energies can be generated by using two closely spaced resonances. For this purpose, we tuned the laser to an energy at which double peaks can be excited simultaneously (peaks B and C in Fig. 27b). The quantum interference between these two resonances results in a slow oscillation superimposed on the rapid oscillation, as shown in Fig. 29. This result implies that the resonant lines observed in near-field PLE are coherently coupled to the ground state. Therefore the result opens up possibilities for coherent optical control of carrier wave functions by use of phonon resonances in SAQDs.

4.5 Spin Relaxation

In addition to energy and momentum relaxation, spin relaxation is an important relaxation process for excited carriers in a semiconductor. In higher-dimensional structures (i.e., bulk, quantum well), spin relaxation is a fast process because of the presence of the continuum states, in which efficient elastic/inelastic scattering can lead to fast spin-flip transitions. In contrast,

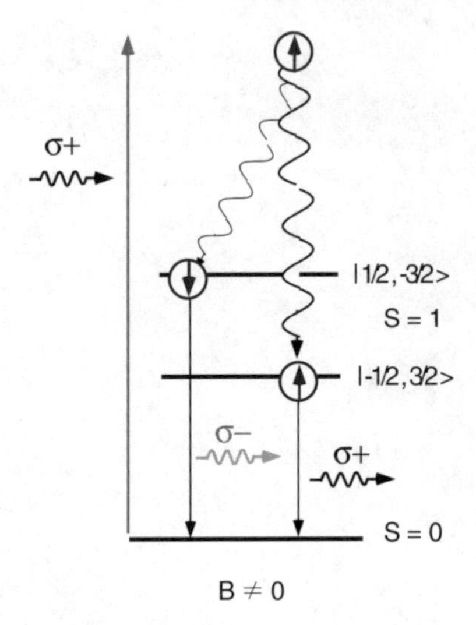

Fig. 30. Schematic diagram of spin relaxation in a magnetic field

discrete DOS in quantum dots restricts possible scattering processes in carrier relaxation. Spin relaxation thus yields useful insight into the physics of excited carriers in 0D systems [40]. In a magnetic field, the ground state splits into two optically allowed states, which exhibit different energy and polarity. Therefore, we can investigate the spin relaxation of excited carriers using selective spin excitations.

Selective excitation of different spin states can be realized by polarization-dependent spectroscopy. Considering the optical transitions between the conduction band ($|\pm 1/2\rangle$) and the heavy-hole valance bands($|\pm 3/2\rangle$), for an optically excited electron–hole pair there are two allowed states: $|\text{e,hh}\rangle = |+1/2, -3/2\rangle$ and $|-1/2, +3/2\rangle$. Recombination of $|+1/2, -3/2\rangle$ and $|-1/2, +3/2\rangle$ generates left-hand circular (σ^-) light and right-hand circular (σ^+) light, respectively. Therefore spectroscopy with polarization discrimination provides information about the spin states. In addition, excitation energy dependence enables us to investigate spin relaxation. For example, examining excited carriers with σ^+ light using a quarter-wave plate in an appropriate state, the carriers relax to the ground state with or without changing their spin polarities (see Fig. 30). They then emit light of either circular polarization as they recombine. We can separate the luminescence of different polarities by another wave plate and polarizer, and observe the spin-flip transition from its polarization.

Figure 31 shows luminescence spectra for different polarization geometries at 8 T. In this measurement, we use InAs/GaAs SAQDs grown by molecular beam epitaxy. But their optical properties are almost the same as we discussed in the previous sections. The carriers are excited with circularly (Fig. 31a–b) or linearly polarized light (Fig. 31c), and the resulting luminescence polarizations are measured by turning a variable wave plate. The

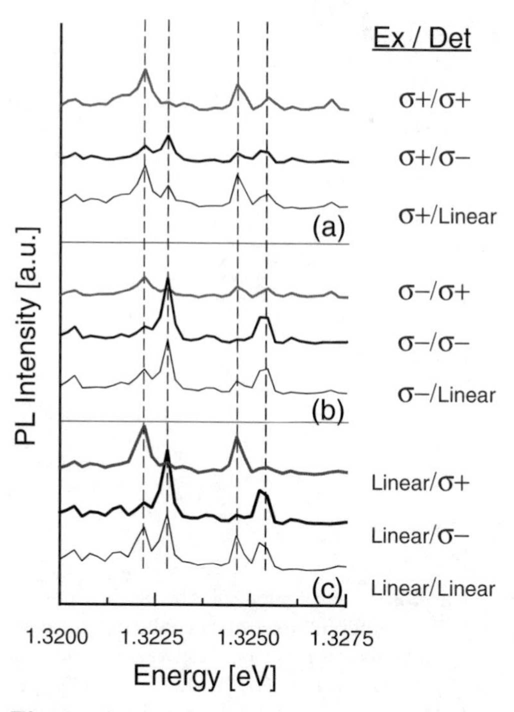

Fig. 31. Luminescence spectra for all polarization geometries at 8 T. Luminescence is dispersed by a single 250-mm monochromator. Two pairs of Zeeman-split line pairs can be seen. The excitation energy is 1.409 eV [41]

excitation energy is set to be 1.4089 eV, corresponding to a relaxation energy of about 80 meV. Since the absorption band edge of the WL is around 1.42 eV, an excitation energy of 1.4089 eV results in carriers being excited in dot states. When carriers are excited with linearly polarized light, the opposite polarities of the split lines are clearly shown (see Fig. 31c). When the excitation and detection polarizations are crossed, the intensities of the split peaks are small compared to when polarizations are the same. There still remain small peaks in crossed polarization due to imperfect polarization selectivity of the experimental setup or to spin relaxation within the recombination.

Figures 32a–d show PL at 8 T with different excitation polarizations at various excitation energies. The splitting energy is estimated to be about 0.6 meV, where spin relaxation within a recombination is sufficiently suppressed. In Fig. 32a, the split peaks show the same relative intensities in each excitation configuration. In contrast, when the laser energy is tuned to lower energies, for each circular excitation polarization one of the two peaks is suppressed, due to selective excitation. In Fig. 32d, this effect is clearly seen. As discussed in Sect. 4.2, the DOS of SAQD exhibit a 2D-like continuum in

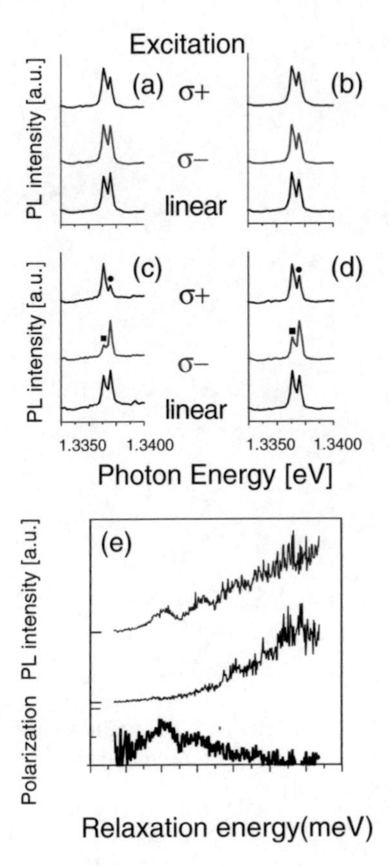

Fig. 32. Comparison of the split lines at 8 T in different excitation polarization configurations and various excitation energies E_{ex} = (**a**) 1.476, (**b**) 1.442, (**c**) 1.425, (**d**) 1.409 eV. (**e**) Corresponding PLE spectra, which are recorded with excitation/detection $= \sigma^-/\sigma^-$ (*top*) and σ^-/σ^+ polarizations. (*Bottom*) Polarization calculated from the upper two spectra [41]

the region above $E_{rel} \sim 50$ meV. In Fig. 32d, where the excitation energy corresponds to the lower end of the continuum, the carriers are excited in a dot state. Therefore suppressed spin relaxation is observed.

On the other hand, in Fig. 32a the carriers are excited in the WL, in which fast spin relaxation can occur by analogy with higher dimensional structures. In Fig. 32c, polarized PL also shows suppressed spin-flip transitions but slightly less poralization dependence than in (d). In order to investigate the changes of spin relaxation more precisely, we measure PLE spectra in the different polarization configurations with σ^- excitation. The WL absorption edge is at a relaxation energy of around 100 meV in these spectra. In the upper two spectra, the detection energy and polarization are set to those of the higher and lower energy peaks, respectively. The magnitude of the polarization is given by $|I^+ - I^-|/|I^+ + I^-|$, where I^+ (I^-) is the PL intensity for detection with the same (opposite) polarity. The calculated polarization is shown at the bottom in Fig. 32e. It can be seen that when the excitation energy is decreased, the polarization is increased. The PLE peaks at 70 meV and 80 meV observed in the σ^-/σ^- configuration are not seen in the σ^-/σ^+

configuration, where the polarization of excitation and detection is opposite. As the upper discrete states of the quantum dots become close in energy to the 2D WL, strong interaction between these states will occur, thus increasing the probability of scattering. Moreover, the degree of polarization changes continuously from the energy of the WL absorption edge to that of the dot states. In Sect. 4.2, we attribute continuum states observed in near-field PLE to gradual crossover from 0D to 2D in SAQDs. This is the very idea that explains the observed continuous change of spin polarization.

5 Conclusion

This chapter describes optical properties of self-assembled quantum dots (SAQDs) characterized by both near-field and far-field spectroscopy. By comparison with these two sprectrocopic results, we can consider the electron–phonon interactions and relaxation mechanism in SAQDs. In far-field spectroscopy, luminescence with broad linewidth including inhomogeneous ensembles provides information on the distribution of individual dots. In contrast, near-field spectra, in which the reduction of inhomogeneous broadening results in an extremely narrow emission linewidth, allows us to probe the precise optical properties of individual SAQDs. We have carried out several experimental studies involving luminescence, luminescence excitation, and coherent spectroscopy. The experiments successfully suggest that excited carriers strongly interact with localized phonons due to the loss of long-range order of the lattice in SAQDs. The results also provide detailed insight into the efficient relaxation process in SAQDs: carriers can relax within continuum states quickly and make transitions to the excitonic ground state by phonon scattering before PL emission. We believe that our findings help to clear up the confusion in this research field, and will stimulate further work, not only experimentally but also theoretically, to investigate relaxation mechanism in 0D structures.

Acknowledgements

The authors wish to express appreciation for helpful support from Prof. M. Ohts and Dr. S. Mononobe in the preparation of the fiber probe. We also would like to thank many colleagues: M. Nishioka, S. Ishida, Dr. K. Suzuki, L. Finger, S. Shinomori, K. Kako, Dr. J. Harris, T. Sugimoto, and Dr. T. Someya.

References

1. Y. Arakawa, H. Sakaki: Appl. Phys. Lett. **40**, 939 (1982)
2. K. Mukai, Y. Nakata, K. Otsubo, M. Sugawara, N. Yokoyama, H. Ishikawa: IEEE J. Quantum Electron. **36**, 472 (2000); F. Heinrichsdorff, Ch. Ribbat, M. Grundmann, D. Bimberg: Appl. Phys. Lett. **76**, 556 (2000)

3. J.Y. Marzin, J.M. Gerard, A. Izral, D. Barrier, G. Bastard: Phys. Rev. Lett. **73**, 716 (1994)
4. J. Oshinowo, M. Nishioka, S. Ishida, Y. Arakawa: Appl. Phys. Lett. **65**, 1421 (1994)
5. H. Benisty, S.M. Sotomayor-Torres, C. Weisbuch: Phys. Rev. B **44**, R10945 (1991); U. Bockelmann: Phys. Rev. B **48**, 17637 (1993)
6. U. Bockelmann, T. Egler: Phys. Rev. B **46**, 15574 (1992); Al.L. Efros, V.A. Karchenko, M. Rosen: Solid State Commun. **93**, 281 (1995)
7. R. Heitz, M. Grundmann, N.N. Ledentsov, L. Eckey, M. Veit, D. Bimberg, V.M. Ustinov, A.Y. Egorov, A.E. Zhukov, P.S. Kop'ev, Z.I. Alferov: Appl. Phys. Lett. **68**, 361 (1996); R. Heitz, M. Veit, N.N. Ledentsov, A. Hoffmann, D. Bimberg, V.M. Ustinov, P.S. Kop'ev, Z.I. Alferov: Phys. Rev. B **56**, 10435 (1997)
8. T. Inoshita, H. Sakaki: Phys. Rev. B **46**, 7260 (1992)
9. X.Q. Li, H. Nakayama, Y. Arakawa: Phys. Rev. B **59**, 5069 (1999)
10. S. Hameau, Y. Guldner, O. Verzelen, R. Ferreira, G. Bastard, J. Zeman, A. Lemaitre, J.M. Gerard: Phys. Rev. Lett. **83**, 4152 (1999)
11. A. Lemaître, A.D. Ashmore, J.J. Finley, D.J. Mowbray, M.S. Skolnick, M. Hopkinson, T.F. Krauss: Phys. Rev. B **63**, 161309 (2001); R. Heitz, I. Mukhametzhanov, O. Stier, A. Madhukar, D. Bimberg: Phys. Rev. Lett. **83**, 4654 (1999); M. Bissiri, G. Baldassarri, Höer von Högersthal, A.S. Bhatti, M. Capizzi, A. Frova, P. Frigeri, S. Franchi: Phys. Rev. B **62**, 4642 (2000)
12. D. Leonard, M. Krishnamurthy, C.M. Reaves, S.P. Denbaars, P.M. Petroff: Appl. Phys. Lett. **63**, 3203 (1993)
13. Y.A. Pusep, G. Zanelatto, S.W. da Silva, J.C. Galzerani, P.P. Gonzalez-Borrero, A.I. Toropov, P. Basmaji: Phys. Rev. B **58**, R1770 (1998); J. Appl. Phys. **86**, 4387 (1999)
14. M. Ichimura, A. Usami, M. Tabichi, A. Sasaki: Phys. Rev. B **51**, 13231 (1995)
15. J. Groenen, G. Landa, R. Carles, P.S. Pizani, M. Gendry: J. Appl. Phys. **82**, 803 (1997)
16. Y. Toda, S. Shinomori, K. Suzuki, M. Nishioka, Y. Arakawa: Solid State Electron. **42**, 1083 (1998)
17. D.W. Pohl: Surface Science **181**, 174 (1987)
18. E. Betzig, J.K. Trautman: Science **257**, 189 (1992)
19. H. Ghaemi, C. Cates, B.B. Goldberg: Ultramicroscopy **57**, 165 (1995)
20. T. Saiki, K. Matsuda: Appl. Phys. Lett. **74**, 2773 (1999)
21. S. Mononobe, T. Saiki, T. Suzuki, S. Koshihahara, M. Ohtsu: Opt. Commun. **146**, 45 (1998)
22. T. Sugimoto, Y. Toda, S. Ishida, M. Nishioka, Y. Arakawa: Technical Digest of *Quantum Electronics and Laser Science Conference, San Francisco, May 7–May 12, 2000* pp. 75–76
23. E. Dekel, D. Gershoni, E. Ehrenfreund, D. Spektor, J.M. Garcia, P.M. Petroff: Phys. Rev. Lett. **80**, 4991 (1998)
24. L. Landin, M.S. Miller, M.-E. Pistol, C.E. Pryor, L. Samuelson: Science **280**, 262 (1998)
25. D. Gammon, E.S. Snow, B.V. Shanabrook, D.S. Katzer, D. Park: Science **273**, 87 (1996)
26. K. Matsuda, K. Ikeda, T. Saiki, H. Tsuchiya, H. Saito, K. Nishi: Phys. Rev. B **63**, R121304 (2001)

27. T. Takagahara: Phys. Rev. B **60**, 2638 (1999)
28. Y. Toda, S. Shinomori, K. Suzuki, Y. Arakawa: Appl. Phys. Lett. **73**, 517 (1998)
29. M. Bayer, A. Kuther, F. Schaäfer, J.P. Reithmaier, A. Forchel: Phys. Rev. B **60**, R8481 (1999)
30. Y. Toda, O. Moriwaki, M. Nishioka, Y. Arakawa: Phys. Rev. Lett. **82**, 4114 (1999)
31. M. Sopanen, H. Lipsanen, J. Ahopelto: Appl. Phys. Lett. **66**, 2364 (1995)
32. F. Findeis, A. Zrenner, G. Böhm, G. Abstreiter: Phys. Rev. B **61** R10579 (2000)
33. J.J. Finley, A.D. Ashmore, A. Lemare, D.J. Mowbray, M.S. Skolnick, I.E. Itskevich, P.A. Maksym, M. Hopkinson, T.F. Krauss: Phys. Rev. B **63**, 73307 (2001)
34. H. Htoon, D. Kulik, O. Baklenov, A.L. Holmes, Jr., T. Takagahara, C.K. Shih: Phys. Rev. B **63**, R241303 (2001)
35. M. Bayer, A. Schmidt, A. Forchel, F. Faller, T.L. Reinecke, P.A. Knipp, A.A. Dremin, V.D. Kulakovskii: Phys. Rev. Lett. **74**, 3439 (1995)
36. W. Heller, U. Bockelmann: Phys. Rev. B **55**, 4871 (1997)
37. M. Bayer, P. Hawrylak, K. Hinzer, S. Fafard, M. Korkusinski, Z.R. Wasilewski, O. Stern, A. Forchel: Science **291**, 451 (2001)
38. N.H. Bonadeo, J. Erland, D. Gammon, D. Park, D.S. Katzer, D.G. Steel: Science **282** 1473 (1998)
39. Y. Toda, T. Sugimoto, M. Nishioka, Y. Arakawa: Appl. Phys. Lett. **76**, 3887 (2000)
40. H. Kamada, H. Gotoh, H. Ando, J. Temmyo, T. Tamamura: Phys. Rev. B **60** 5791 (1999)
41. Y. Toda, S. Shinomori, K. Suzuki, Y. Arakawa: Phys. Rev. B **58**, R10147 (1998)

Quantum Theoretical Approach to Optical Near-Fields and Some Related Applications

K. Kobayashi, S. Sangu, and M. Ohtsu

1 Introduction

Several theoretical approaches, different from their viewpoints to optical near-field problems, have been proposed for a decade. The essential points in difference are related to what the optical near-field interaction is, or how materials respond to light confined in a small area less than the wavelength. The so-called classical approach that is based on the macroscopic Maxwell equations extends the theory to describe electromagnetic phenomena taking place on a-sub wavelength or even on a nanometer scale, while matter response to the electromagnetic field is represented by a macroscopic refractive index or a dielectric constant. The main concern in this approach is to find field distributions around the material system after solving the macroscopic Maxwell equations. Note that the electromagnetic fields in or near the material, are averaged over a large dimension, but still smaller than the wavelength, satisfy the macroscopic Maxwell equations, and that the "bare" electromagnetic fields on a nanometer scale should be governed by the microscopic Maxwell equations [1]. A number of analytical and numerical methods have been reported in order to obtain the relevant electromagnetic fields [2,3].

The semiclassical approach employs the Schrödinger equation to handle the material (electron) response by focusing on the internal energy structure of the material system, though electromagnetic fields are still not quantized. According to the electronic state of the material system, the susceptibility at an arbitrary point of the system is obtained, and thus the local electromagnetic field combined with the susceptibility determines the polarization locally induced in the system. Macroscopic electromagnetic fields are then given by the macroscopic Maxwell equations with the induced polarization as a source term. We have a final solution of the problem after the induced polarization and the macroscopic electromagnetic fields are made self-consistent by iteration. It should be noted that the induced polarization at an arbitrary position is affected by the local electromagnetic fields at other positions, as well as the same position, because the susceptibility is nonlocal [4]. This nonlocality becomes important when the electron correlation length of the material system is comparable to the spatial averaging length of the microscopic electromagnetic fields [5,6]. Recent advances in optical near-field experiments may reveal the nature of these phenomena.

The semiclassical approach implies the importance of the coupling between the matter and the electromagnetic fields on a sub-wavelength scale, and of a microscopic formulation. If electromagnetic fields are quantized as well, we may also expect the analogous discussion provoked in the realm of quantum optics to open up a new field. However, it is difficult to develop a quantized nonlocal theory, and the difficulty forces us to adopt another standpoint. In this article we avoid it, and will develop a new formulation within a quantum-mechanical framework, putting matter excitations on an equal footing with photons.

1.1 Basic Idea and Massive Virtual Photon Model

There is no way to obtain all the energy states of all the microscopic constituents in a complicated system, or to describe time evolution of the states. In most cases, one prefers to have any information about characteristic states or observables, instead of all the exact solutions. Projection operator methods and elementary excitation modes have been developed in this context, and will be discussed in the following sections.

Let us consider an optical near-field system from this viewpoint. It consists of a macroscopic matter system larger than the wavelength of incident light (light source, fiber probe, prism substrate, optical detector, etc.) and a nanometric aperture or a protrusion of the probe tip as well as a nanometric matter system (sample material). Those systems are interacting with one another via optical near fields. Because of the huge number of degrees of freedom and complexity, it is impossible to obtain exact solutions of the system from the viewpoint of an atomic many-body system that consists of electrons and nuclei interacting with photons. We are not interested in exact solutions of such a many-body system. Instead we are willing to know answers of fundamental optical near-field problems: What kind of interaction is exerted between the probe tip and sample? How is the interaction manifested? Why does the lateral resolution exceed the diffraction limit of the incident light? What kind of change should be made if we remove the assumption that the matter system is isolated from the other systems when we study the interaction between propagating light and matter.

The massive virtual photon model [7–9] has been proposed in order to examine such fundamental issues and promote specific advanced experiments. The model assumes that the effective optical near-field interaction between a probe and sample system is described by the Yukawa potential, as a result of mediation of a massive virtual photon, not the usual massless photon. It indicates that the interaction range is finite, and that it is active only if the probe and the sample are closely located. According to the Heisenberg uncertainty relation, a process where the energy conservation law is not satisfied can occur in a short period compared with the macroscopic time required for measurement. The massive virtual photon is the quantum associated with such a process, and is a so-called "dressed photon" whose mass is attributed to the coupling with matter excitation, as will be discussed later.

It is well known that a scalar meson with effective mass m_{eff} satisfies the Klein–Gordon equation [10]

$$\left[\Delta - \left(\frac{m_{\mathrm{eff}} c}{\hbar}\right)^2\right] \phi(\boldsymbol{r}) = 0 \,, \tag{1}$$

and that it has a solution

$$\phi(r) = \frac{\exp\left(-\frac{m_{\mathrm{eff}} c}{\hbar} r\right)}{r} \,. \tag{2}$$

Here $\hbar$ and c are the Planck constant divided by 2π and the speed of light in vacuum, respectively. By way of comparison we note that an electric field $\boldsymbol{E}$ as a function of an angular frequency ω and position $\boldsymbol{r}$ is described by the Helmholtz equation [1]

$$\left[\Delta + \left(\frac{\omega}{c}\right)^2\right] \boldsymbol{E}(\boldsymbol{r}, \omega) = -\frac{1}{\varepsilon_0}\left(\frac{\omega}{c}\right)^2 \boldsymbol{P}(\boldsymbol{r}, \omega) \,. \tag{3}$$

One can infer that the electric field given by (3) is transformed into the Klein–Gordon field after the induced polarization $\boldsymbol{P}(\boldsymbol{r}, \omega)$ is renormalized into the effective mass as a result of many-body effects.

Using the projection operator method and elementary excitation modes, we formulate the idea of the virtual photon model and the optical near-field problems in the following sections. At the same time, some examples related to atom photonics and nano photonics are discussed as an application of the formulation. Before getting into the detail of the formulation, the projection operator method is briefly outlined. Starting with the definition of the projection operator, some useful properties are discussed in the next section. Then the effective operator and effective interaction are derived in Sect. 3. In Sect. 4 the "bare interaction" is described in two ways in order to obtain more explicit expressions of the effective interaction. We introduce elementary excitation modes to describe electronic polarization in macroscopic matter systems in Sect. 5; in particular, the exciton polariton concept is discussed in some detail. Section 6 is devoted to the main part of the formulation. As a related application of our approach, we address in Sect. 7 basic problems of single-atom manipulation with a near-field optical probe, as well as conventional problems on optical near-field microscopy. Finally we discuss prospects for this study.

2 Projection Operator Method

In this section we discuss what the projection operator is and what kind of properties it has.

2.1 Definition of the Projection Operator

When one considers an interacting system such as a system consisting of an isolated quantum system and electromagnetic field, the Hamiltonian operator $\hat{H}$ for the total system is represented as a sum of $\hat{H}_0$ and $\hat{V}$ as

$$\hat{H} = \hat{H}_0 + \hat{V} , \tag{4}$$

where $\hat{H}_0$ describes the isolated system while $\hat{V}$ represents the interaction. If eigenstates and eigenvalues of the Hamiltonian $\hat{H}$ are written as $|\psi_j\rangle$ and E_j, respectively, the following Schrödinger equation holds:

$$\hat{H} |\psi_j\rangle = E_j |\psi_j\rangle . \tag{5}$$

Here the suffix j is used to specify the quantum numbers that distinguish each eigenstate. In a similar way we denote eigenstates of the Hamiltonian $\hat{H}_0$ as $|\phi_j\rangle$. Then we define the projection operator P as

$$P = \sum_{j=1}^{N} |\phi_j\rangle \langle\phi_j| , \tag{6}$$

where N is an arbitrary integer, but it is preferably a small number in practice. Operating with the projection operator on an arbitrary state $|\psi\rangle$, we obtain

$$P |\psi\rangle = \sum_{j=1}^{N} |\phi_j\rangle \langle\phi_j| \psi\rangle . \tag{7}$$

From this relation it follows that the projection operator transforms an arbitrary state $|\psi\rangle$ into the P space spanned by the eigenstates $|\phi_j\rangle$. We have defined the projection operator based on stationary states of the Schrödinger equation. Readers who are interested in the time- dependent approach to the projection operator method are referred to [11–16].

2.2 Properties of the Projection Operator

Using the projection operator P, we can derive an effective operator $\hat{O}_{\text{eff}}$ for an arbitrary operator $\hat{O}$ that corresponds to a physical observable. In order to perform the calculation, we begin with some elementary and useful properties of the projection operator. Since the eigenstate $|\phi_j\rangle$ is orthonormalized, the projection operator P satisfies

$$P = P^{\dagger} , \quad P^2 = P . \tag{8}$$

The complementary operator Q given by

$$Q = 1 - P \tag{9}$$

satisfies

$$Q = Q^{\dagger}, \quad Q^2 = Q . \tag{10}$$

Any state in the P space is orthogonal to any state in the Q space, and thus we have

$$PQ = QP = 0 . \tag{11}$$

Note that $|\phi_j\rangle$ is an eigenstate of $\hat{H}_0$, so the commutator of the projection operator and $\hat{H}_0$ vanishes:

$$\left[P, \hat{H}_0\right] = P\hat{H}_0 - \hat{H}_0 P = 0 , \tag{12a}$$

$$\left[Q, \hat{H}_0\right] = Q\hat{H}_0 - \hat{H}_0 Q = 0 . \tag{12b}$$

3 Effective Operator and Effective Interaction

Dividing the eigenstates $|\psi_j\rangle$ into two groups, we define $\left|\psi_j^{(1)}\right\rangle$ in the P space and $\left|\psi_j^{(2)}\right\rangle$ in the Q space as follows:

$$\left|\psi_j^{(1)}\right\rangle = P\left|\psi_j\right\rangle , \tag{13a}$$

$$\left|\psi_j^{(2)}\right\rangle = Q\left|\psi_j\right\rangle . \tag{13b}$$

Then we obtain a set of equations for $\left|\psi_j^{(1)}\right\rangle$ and $\left|\psi_j^{(2)}\right\rangle$. From (9), we have

$$|\psi_j\rangle = P\left|\psi_j^{(1)}\right\rangle + Q\left|\psi_j^{(2)}\right\rangle , \tag{14}$$

and from (4) and (5)

$$\left(E_j - \hat{H}_0\right)|\psi_j\rangle = \hat{V}|\psi_j\rangle \tag{15}$$

is derived. Inserting (14) into (15), we obtain

$$\left(E_j - \hat{H}_0\right) P\left|\psi_j^{(1)}\right\rangle + \left(E_j - \hat{H}_0\right) Q\left|\psi_j^{(2)}\right\rangle = \hat{V}P\left|\psi_j^{(1)}\right\rangle + \hat{V}Q\left|\psi_j^{(2)}\right\rangle . \tag{16}$$

Operating with P from the left on (16) and using (8), (11), (12a) and (12b), we obtain

$$\left(E_j - \hat{H}_0\right) P\left|\psi_j^{(1)}\right\rangle = P\hat{V}P\left|\psi_j^{(1)}\right\rangle + P\hat{V}Q\left|\psi_j^{(2)}\right\rangle . \tag{17}$$

Similarly, applying Q from the left on (16), we have

$$\left(E_j - \hat{H}_0\right) Q \left|\psi_j^{(2)}\right\rangle = Q\hat{V}P \left|\psi_j^{(1)}\right\rangle + Q\hat{V}Q \left|\psi_j^{(2)}\right\rangle . \tag{18}$$

From (18), it is possible to formally express $Q\left|\psi_j^{(2)}\right\rangle$ in terms of $P\left|\psi_j^{(1)}\right\rangle$ as

$$\begin{aligned} Q\left|\psi_j^{(2)}\right\rangle &= \left(E_j - \hat{H}_0 - Q\hat{V}\right)^{-1} Q\hat{V}P\left|\psi_j^{(1)}\right\rangle \\ &= \left\{\left(E_j - \hat{H}_0\right)\left[1 - \left(E_j - \hat{H}_0\right)^{-1} Q\hat{V}\right]\right\}^{-1} Q\hat{V}P\left|\psi_j^{(1)}\right\rangle \\ &= \hat{J}\left(E_j - \hat{H}_0\right)^{-1} Q\hat{V}P\left|\psi_j^{(1)}\right\rangle , \end{aligned} \tag{19}$$

where the operator $\hat{J}$ is defined as

$$\hat{J} = \left[1 - \left(E_j - \hat{H}_0\right)^{-1} Q\hat{V}\right]^{-1} . \tag{20}$$

Substituting (19) into the second term on the right- hand side of (17), one obtains the equation for $P\left|\psi_j^{(1)}\right\rangle$:

$$\begin{aligned} \left(E_j - \hat{H}_0\right) P\left|\psi_j^{(1)}\right\rangle &= P\hat{V}P\left|\psi_j^{(1)}\right\rangle + P\hat{V}\hat{J}\left(E_j - \hat{H}_0\right)^{-1} Q\hat{V}P\left|\psi_j^{(1)}\right\rangle \\ &= P\hat{V}\hat{J}\left\{\hat{J}^{-1} + \left(E_j - \hat{H}_0\right)^{-1} Q\hat{V}\right\} P\left|\psi_j^{(1)}\right\rangle \\ &= P\hat{V}\hat{J}P\left|\psi_j^{(1)}\right\rangle . \end{aligned} \tag{21}$$

On the other hand, if (19) is inserted into (14), then we have

$$\begin{aligned} |\psi_j\rangle &= P\left|\psi_j^{(1)}\right\rangle + \hat{J}\left(E_j - \hat{H}_0\right)^{-1} Q\hat{V}P\left|\psi_j^{(1)}\right\rangle \\ &= \hat{J}\left\{\hat{J}^{-1} + \left(E_j - \hat{H}_0\right)^{-1} Q\hat{V}\right\} P\left|\psi_j^{(1)}\right\rangle \\ &= \hat{J}P\left|\psi_j^{(1)}\right\rangle . \end{aligned} \tag{22}$$

Noticing the normalization condition for $|\psi_j\rangle$ and $\left|\psi_j^{(1)}\right\rangle$, it is possible to rewrite (22):

$$|\psi_j\rangle = \hat{J}P\left(P\hat{J}^\dagger\hat{J}P\right)^{-1/2}\left|\psi_j^{(1)}\right\rangle . \tag{23}$$

Since $|\psi_j\rangle$ has been expressed in terms of $\left|\psi_j^{(1)}\right\rangle$, one can obtain an effective operator $\hat{O}_{\text{eff}}$ from the following relation:

$$\langle\psi_i|\,\hat{O}\,|\psi_j\rangle = \left\langle\psi_i^{(1)}\right|\hat{O}_{\text{eff}}\left|\psi_j^{(1)}\right\rangle . \tag{24}$$

Substituting (23) on the left-hand side of (24) and comparing it with the right-hand side, we obtain [17–19]

$$\hat{O}_{\text{eff}} = \left(P\hat{J}^\dagger \hat{J} P\right)^{-1/2} \left(P\hat{J}^\dagger \hat{O} \hat{J} P\right) \left(P\hat{J}^\dagger \hat{J} P\right)^{-1/2} . \qquad (25)$$

If $\hat{V}$ is taken as $\hat{O}$, then the effective interaction operator $\hat{V}_{\text{eff}}$ is written as

$$\hat{V}_{\text{eff}} = \left(P\hat{J}^\dagger \hat{J} P\right)^{-1/2} \left(P\hat{J}^\dagger \hat{V} \hat{J} P\right) \left(P\hat{J}^\dagger \hat{J} P\right)^{-1/2} . \qquad (26)$$

This is what we are searching for. As expected, $\hat{V}_{\text{eff}}$ is defined so as to operate only on states in the P space. Once the bare interaction $\hat{V}$ is given, it only remains to obtain the unknown operator $\hat{J}$.

3.1 Equation for the Operator $\hat{J}$ and Its Approximate Solution

In order to obtain an explicit form of the operator $\hat{J}$, let us consider the operator $\left[\hat{J}, \hat{H}_0\right] P$ and apply it to the state $|\psi_j\rangle$. This yields

$$\begin{aligned} \left[\hat{J}, \hat{H}_0\right] P |\psi_j\rangle &= \left(\hat{J}\hat{H}_0 - \hat{H}_0\hat{J}\right) P |\psi_j\rangle \\ &= \left\{\left(E_j - \hat{H}_0\right)\hat{J} - \hat{J}\left(E_j - \hat{H}_0\right)\right\} P |\psi_j\rangle . \end{aligned} \qquad (27)$$

We replace the first term $\left(E_j - \hat{H}_0\right)$ in (27) by $\hat{V}$, which gives

$$\left[\hat{J}, \hat{H}_0\right] P |\psi_j\rangle = \hat{V}\hat{J}P |\psi_j\rangle - \hat{J}\left(E_j - \hat{H}_0\right) P |\psi_j\rangle . \qquad (28)$$

Using (17) and (19), the second term of (28) can be rewritten as

$$\begin{aligned} \hat{J}\left(E_j - \hat{H}_0\right) P |\psi_j\rangle &= \hat{J}\left(E_j - \hat{H}_0\right) P \left|\psi_j^{(1)}\right\rangle \\ &= \hat{J}\left\{P\hat{V}P\left|\psi_j^{(1)}\right\rangle + P\hat{V}Q\left|\psi_j^{(2)}\right\rangle\right\} \\ &= \hat{J}P\hat{V}\left\{P\left|\psi_j^{(1)}\right\rangle + \hat{J}\left(E_j - \hat{H}_0\right)^{-1} Q\hat{V}P\left|\psi_j^{(1)}\right\rangle\right\} \\ &= \hat{J}P\hat{V}\hat{J}\left\{\hat{J}^{-1} + \left(E_j - \hat{H}_0\right)^{-1} Q\hat{V}\right\} P\left|\psi_j^{(1)}\right\rangle . \end{aligned} \qquad (29)$$

Making use of (20) and noting that $P\left|\psi_j^{(1)}\right\rangle = P|\psi_j\rangle$, we can rewrite (28) as

$$\left[\hat{J}, \hat{H}_0\right] P |\psi_j\rangle = \hat{V}\hat{J}P |\psi_j\rangle - \hat{J}P\hat{V}\hat{J}P |\psi_j\rangle . \qquad (30)$$

Therefore we have for the operator $\hat{J}$

$$\left[\hat{J}, \hat{H}_0\right] P = \hat{V}\hat{J}P - \hat{J}P\hat{V}\hat{J}P \,, \tag{31}$$

where all operators involved are known except $\hat{J}$.

In order to solve (31) perturbatively, let us assume

$$\hat{J} = \sum_{n=0}^{\infty} g^n \hat{J}^{(n)} \,, \tag{32}$$

where the nth term $\hat{J}^{(n)}$ contains n $\hat{V}$s and $\hat{J}^{(0)} = P$. Substituting (32) into (31) and equating terms of order g^n on the two sides, we successively obtain $\hat{J}^{(1)}$, $\hat{J}^{(2)}$, $\cdots$, $\hat{J}^{(n)}$. For example, noting the identity

$$\begin{aligned} Q\left[\hat{J}^{(1)}, \hat{H}_0\right] P &= Q\hat{V}\hat{J}^{(0)}P - Q\hat{J}^{(0)}P\hat{V}\hat{J}^{(0)}P \\ &= Q\hat{V}P \,, \end{aligned} \tag{33}$$

we take the matrix element of (33) with $|\psi_j\rangle$;

$$\begin{aligned} \langle\psi_i| Q\left[\hat{J}^{(1)}, \hat{H}_0\right] P |\psi_j\rangle &= \langle\psi_i| Q\hat{J}^{(1)}\left(E_P^0 - E_Q^0\right) P |\psi_j\rangle \\ &= \langle\psi_i| Q\hat{V}P |\psi_j\rangle \,, \end{aligned} \tag{34}$$

where we used the eigenvalues E_P^0 and E_Q^0 of the Hamiltonian $\hat{H}_0$ in the P space and Q space, respectively. From (34),

$$\hat{J}^{(1)} = Q\hat{V}\left(E_P^0 - E_Q^0\right)^{-1} P \,. \tag{35}$$

Higher orders of $\hat{J}^{(n)}$ are successively given in a similar way.

3.2 Effective Interaction Operator in an Approximation

After inserting the perturbative solution discussed above into the operator $\hat{J}$, it is possible to approximately express $\hat{V}_{\text{eff}}$ in (26), which operates only on states in the P space. Using $\hat{J}^{(0)}$, one can obtain $\hat{V}_{\text{eff}}$ in lowest order as

$$\hat{V}_{\text{eff}} = P\hat{V}P \,, \tag{36}$$

which is equivalent to the so-called "bare interaction" because all effects from the Q space are neglected. The multipole Hamiltonian, which will be discussed as one of the bare interactions in the next section, cannot contribute to (36) if transverse photon states are not included in the P space. It is therefore necessary to employ $\hat{J}^{(1)}$ at least, in order to include Q- space effects. Then we have

$$\hat{V}_{\text{eff}} = 2P\hat{V}Q\left(E_P^0 - E_Q^0\right)^{-1} Q\hat{V}P \,, \tag{37}$$

and in the following we discuss the optical near-field interaction based on the formula (37). Whenever one improves the approximation of $\hat{J}$, one can examine the higher-order effects originating in the Q space. These procedures correspond to how to obtain many-body Green's functions for matter systems, or Green's functions for photons "dressed with matter excitations" [20].

4 Electromagnetic Interaction with Matter: Minimal-Coupling and Multipole Hamiltonians

The "bare interaction" must be specified in order to obtain a more explicit expression of the effective interaction. There are two ways to describe the interaction between an electromagnetic field and a charged particle. One is to use the minimal-coupling Hamiltonian, and the other is to employ the multipole Hamiltonian. These two Hamiltonians are related to each other by a unitary transformation, and there are, in principle, no problems regardless of which is adopted [21–23]. However, it should be noted that the complexity of description depends on each problem: it may be easier for the minimal-coupling Hamiltonian formalism to describe one problem, and more difficult to describe another.

4.1 Minimal-Coupling Hamiltonian

We can derive the minimal-coupling Hamiltonian for a charged particle, i.e., the electromagnetic interaction with a charged particle, by imposing the local gauge invariance on the Hamiltonian describing free particle motion [24]. The minimal coupling Hamiltonian is defined such that the Schrödinger equation is not changed if a wave function $\psi(\boldsymbol{r},t)$ that satisfies the Schrödinger equation is transformed by the phase transformation $\chi(\boldsymbol{r},t)$ to

$$\psi'(\boldsymbol{r},t) = \exp\left[i\chi(\boldsymbol{r},t)\right]\psi(\boldsymbol{r},t) \ , \tag{38}$$

and if vector potential $\boldsymbol{A}(\boldsymbol{r},t)$ and scalar potential $U(\boldsymbol{r},t)$ are transformed by the following gauge transformation [10,21,25,26]:

$$\boldsymbol{A}'(\boldsymbol{r},t) = \boldsymbol{A}(\boldsymbol{r},t) + \frac{\hbar}{e}\nabla\chi(\boldsymbol{r},t) \ , \tag{39a}$$

$$U'(\boldsymbol{r},t) = U(\boldsymbol{r},t) - \frac{\hbar}{e}\frac{\partial}{\partial t}\chi(\boldsymbol{r},t) \ . \tag{39b}$$

Here $\hbar$ and e are the Planck constant divided by 2π and the electric charge of the particle. Let us simply assume the electromagnetic fields to be classical in this section. In order to satisfy the above requirement, it follows that the Hamiltonian must be

$$H' = \frac{1}{2m}\left[\boldsymbol{p} - e\boldsymbol{A}'(\boldsymbol{r},t)\right]^2 + eU'(\boldsymbol{r},t) \ , \tag{40}$$

where the mass and the momentum of the particle are denoted by m and $\boldsymbol{p}$, respectively. For confirmation, (38), (39a), (39b), and (40) are substituted into the Schrödinger equation:

$$i\hbar\frac{\partial}{\partial t}\psi'(\boldsymbol{r},t) = H'\psi'(\boldsymbol{r},t) . \tag{41}$$

Noting that the momentum $\boldsymbol{p}$ must be an operator, $-i\hbar\nabla$, it follows that the left-hand side of (41) reads

$$-\hbar\exp\left[i\chi(\boldsymbol{r},t)\right]\frac{\partial\chi(\boldsymbol{r},t)}{\partial t}\psi(\boldsymbol{r},t)+i\hbar\exp\left[i\chi(\boldsymbol{r},t)\right]\frac{\partial\psi(\boldsymbol{r},t)}{\partial t} . \tag{42}$$

On the other hand, the right-hand side of (41) can be rewritten as

$$\begin{aligned}\frac{1}{2m}&\left[-i\hbar\nabla - e\left\{\boldsymbol{A}(\boldsymbol{r},t)+\frac{\hbar}{e}\nabla\chi(\boldsymbol{r},t)\right\}\right]^2\exp\left[i\chi(\boldsymbol{r},t)\right]\psi(\boldsymbol{r},t)\\ &\qquad+e\left\{U(\boldsymbol{r},t)-\frac{\hbar}{e}\frac{\partial}{\partial t}\chi(\boldsymbol{r},t)\right\}\exp\left[i\chi(\boldsymbol{r},t)\right]\psi(\boldsymbol{r},t)\\ &=\exp\left[i\chi(\boldsymbol{r},t)\right]\left\{\frac{1}{2m}\left[-i\hbar\nabla - e\boldsymbol{A}(\boldsymbol{r},t)\right]^2+eU(\boldsymbol{r},t)\right\}\psi(\boldsymbol{r},t)\\ &\qquad-\hbar\exp\left[i\chi(\boldsymbol{r},t)\right]\frac{\partial\chi(\boldsymbol{r},t)}{\partial t}\psi(\boldsymbol{r},t) .\end{aligned} \tag{43}$$

Therefore we can leave the Schrödinger equation unchanged if

$$i\hbar\frac{\partial}{\partial t}\psi(\boldsymbol{r},t) = H\psi(\boldsymbol{r},t) , \tag{44a}$$

$$H = \frac{1}{2m}\left[-i\hbar\nabla - e\boldsymbol{A}(\boldsymbol{r},t)\right]^2 + eU(\boldsymbol{r},t) . \tag{44b}$$

In other words, the relevant Hamiltonian is obtained by formally adding $eU(\boldsymbol{r},t)$ and replacing $\boldsymbol{p}$ by $\boldsymbol{p}-e\boldsymbol{A}(\boldsymbol{r},t)$ in the Hamiltonian for a free particle. From (44b), the interaction Hamiltonian for the electromagnetic field and the charged particle consists of two parts:

$$H_1 = -\frac{e}{m}\boldsymbol{p}\cdot\boldsymbol{A}(\boldsymbol{r},t) , \tag{45a}$$

$$H_2 = \frac{e^2}{2m}\boldsymbol{A}^2(\boldsymbol{r},t) . \tag{45b}$$

Advantages of this form of the Hamiltonian are that it can easily describe relativistic covariance and is firmly rooted in gauge theory [25,26]. However, it has the disadvantage that exact description including retardation is cumbersome in the Coulomb gauge ($\nabla\cdot\boldsymbol{A}=0$), where the transversality of light is considered to be important in order to handle the interaction between light and matter as a many-body system.

4.2 Multipole Hamiltonian

In this subsection we discuss another form of the light–matter interaction via a unitary transformation of the minimal coupling Hamiltonian, which removes the disadvantages mentioned above. Such Hamiltonian has a simple form without the static Coulomb interaction, and can exactly describe the retardation by exchanging only transverse photons [23].

Let us consider a charged-particle system confined in a microscopic area, and hereinafter call it a molecule. Electric neutrality of the molecule is assumed, and thus it may be an atom or a molecule as a physical entity. In the following, we choose a two-molecule system as an example, and look for an appropriate Hamiltonian. When the wavelength of electromagnetic waves is much greater than the molecular dimension, the vector potential $\boldsymbol{A}(\boldsymbol{R})$ at the center position $\boldsymbol{R}$ of a molecule is the same as $\boldsymbol{A}(\boldsymbol{q})$, independent of the position $\boldsymbol{q}$ of an electric charge in the molecule:

$$\boldsymbol{A}(\boldsymbol{q}) = \boldsymbol{A}(\boldsymbol{R}) \ . \tag{46}$$

From (46), it follows that $\boldsymbol{B} = \nabla \times \boldsymbol{A} = 0$, and thus we can neglect the interaction between the particle and the magnetic field. Moreover the electric dipole interaction, for simplicity, is taken into account, i.e., higher multipoles are neglected. We assume in addition that the electron exchange interaction is also negligible. Then the Lagrangian L for the system, consisting of three parts L_{mol}, L_{rad}, and L_{int}, can be written as

$$L = L_{\mathrm{mol}} + L_{\mathrm{rad}} + L_{\mathrm{int}} \ , \tag{47a}$$

$$L_{\mathrm{mol}} = \sum_{\zeta=1}^{2} \left\{ \sum_{\alpha} \frac{m_\alpha \dot{\boldsymbol{q}}_\alpha^2(\zeta)}{2} - V(\zeta) \right\} , \tag{47b}$$

$$L_{\mathrm{rad}} = \frac{\varepsilon_0}{2} \int \left\{ \dot{\boldsymbol{A}}^2 - c^2 (\nabla \times \boldsymbol{A})^2 \right\} d^3 r \ , \tag{47c}$$

$$L_{\mathrm{int}} = \sum_{\zeta=1}^{2} \sum_{\alpha} e \dot{\boldsymbol{q}}_\alpha(\zeta) \cdot \boldsymbol{A}(\boldsymbol{R}_\zeta) - V_{\mathrm{inter}} \ , \tag{47d}$$

where the index ζ is used to distinguish the molecules 1 and 2, and α is used to specify a charged particle in a molecule. The energy of the charged particles with mass m_α and velocity $\dot{\boldsymbol{q}}_\alpha$ in the Coulomb potential $V(\zeta)$ is denoted by L_{mol}, while L_{rad} represents the energy of the electromagnetic field in free space. The third term in the Lagrangian represents the interaction between the charge and the electromagnetic field and the Coulomb interaction V_{inter} between molecules 1 and 2, which is given by

$$V_{\mathrm{inter}} = \frac{1}{4\pi\varepsilon_0 R^3} \left\{ \boldsymbol{\mu}(1) \cdot \boldsymbol{\mu}(2) - 3 (\boldsymbol{\mu}(1) \cdot \boldsymbol{e}_R)(\boldsymbol{\mu}(2) \cdot \boldsymbol{e}_R) \right\} \ . \tag{48}$$

Here $R = |\boldsymbol{R}| = |\boldsymbol{R}_1 - \boldsymbol{R}_2|$ denotes the distance between the centers of molecules 1 and 2, and $\boldsymbol{e}_R$ is $\boldsymbol{R}/R$, the unit vector in the direction of $\boldsymbol{R}$. The electric dipole moments of molecules 1 and 2 are $\boldsymbol{\mu}(1)$ and $\boldsymbol{\mu}(2)$, respectively.

In order to simplify the interaction Hamiltonian without changing the equations of motion, we carry out the Power–Zienau–Woolley transformation [21] on the original Lagrangian L:

$$L_{\mathrm{mult}} = L - \frac{\mathrm{d}}{\mathrm{d}t} \int \boldsymbol{P}^{\perp}(\boldsymbol{r}) \cdot \boldsymbol{A}(\boldsymbol{r}) \, \mathrm{d}^3 r \,, \tag{49}$$

where $\boldsymbol{P}^{\perp}(\boldsymbol{r})$ is the transverse component of the polarization density $\boldsymbol{P}(\boldsymbol{r})$; this means that transverse photons can only contribute to the second term in (49). The polarization density $\boldsymbol{P}(\boldsymbol{r})$ is

$$\begin{aligned}
\boldsymbol{P}(\boldsymbol{r}) &= \sum_{\zeta,\alpha} e\left(\boldsymbol{q}_\alpha - \boldsymbol{R}_\zeta\right) \int_0^1 \delta\left(\boldsymbol{r} - \boldsymbol{R}_\zeta - \lambda\left(\boldsymbol{q}_\alpha - \boldsymbol{R}_\zeta\right)\right) \mathrm{d}\lambda \\
&= \sum_{\zeta,\alpha} e\left(\boldsymbol{q}_\alpha - \boldsymbol{R}_\zeta\right) \left[1 - \frac{1}{2!}\left\{\left(\boldsymbol{q}_\alpha - \boldsymbol{R}_\zeta\right) \cdot \nabla\right\}\right. \\
&\quad \left. + \frac{1}{3!}\left\{\left(\boldsymbol{q}_\alpha - \boldsymbol{R}_\zeta\right) \cdot \nabla\right\}^2 - \cdots \right] \delta\left(\boldsymbol{r} - \boldsymbol{R}_\zeta\right) \,,
\end{aligned} \tag{50}$$

and only the electric dipole term is retained:

$$\begin{aligned}
\boldsymbol{P}(\boldsymbol{r}) &= \sum_{\zeta,\alpha} e\left(\boldsymbol{q}_\alpha - \boldsymbol{R}_\zeta\right) \delta\left(\boldsymbol{r} - \boldsymbol{R}_\zeta\right) \\
&= \boldsymbol{\mu}(1)\, \delta\left(\boldsymbol{r} - \boldsymbol{R}_1\right) + \boldsymbol{\mu}(2)\, \delta\left(\boldsymbol{r} - \boldsymbol{R}_2\right) \,.
\end{aligned} \tag{51}$$

Note that the current density $\boldsymbol{j}(\boldsymbol{r})$ is

$$\boldsymbol{j}(\boldsymbol{r}) = \sum_{\zeta,\alpha} e \dot{\boldsymbol{q}}_\alpha \delta\left(\boldsymbol{r} - \boldsymbol{R}_\zeta\right) \,, \tag{52}$$

and the transverse component of the current density is related to the transverse component of the polarization density:

$$\boldsymbol{j}^{\perp}(\boldsymbol{r}) = \frac{\mathrm{d}\boldsymbol{P}^{\perp}(\boldsymbol{r})}{\mathrm{d}t} \,. \tag{53}$$

Using (52) and (53), we can rewrite the interaction Lagrangian L_{int} as

$$\begin{aligned}
L_{\mathrm{int}} &= \int \boldsymbol{j}^{\perp}(\boldsymbol{r}) \cdot \boldsymbol{A}(\boldsymbol{r}) \, \mathrm{d}^3 r - V_{\mathrm{inter}} \\
&= \int \frac{\mathrm{d}\boldsymbol{P}^{\perp}(\boldsymbol{r})}{\mathrm{d}t} \cdot \boldsymbol{A}(\boldsymbol{r}) \, \mathrm{d}^3 r - V_{\mathrm{inter}} \,,
\end{aligned} \tag{54}$$

and thus L_{mult} given by (49) becomes

$$\begin{aligned}
L_{\mathrm{mult}} &= L - \int \frac{\mathrm{d}\boldsymbol{P}^{\perp}(\boldsymbol{r})}{\mathrm{d}t} \cdot \boldsymbol{A}(\boldsymbol{r}) \, \mathrm{d}^3 r - \int \boldsymbol{P}^{\perp}(\boldsymbol{r}) \cdot \dot{\boldsymbol{A}}(\boldsymbol{r}) \, \mathrm{d}^3 r \\
&= L_{\mathrm{mol}} + L_{\mathrm{rad}} - \int \boldsymbol{P}^{\perp}(\boldsymbol{r}) \cdot \dot{\boldsymbol{A}}(\boldsymbol{r}) \, \mathrm{d}^3 r - V_{\mathrm{inter}} \,.
\end{aligned} \tag{55}$$

Here we recall the definition of the momentum $\boldsymbol{p}_\alpha$ conjugate to $\boldsymbol{q}_\alpha$, and $\boldsymbol{\Pi}(\boldsymbol{r})$ to $\boldsymbol{A}(\boldsymbol{r})$,

$$\boldsymbol{p}_\alpha = \frac{\partial L_{\text{mult}}}{\partial \dot{\boldsymbol{q}}_\alpha} = \frac{\partial L_{\text{mol}}}{\partial \dot{\boldsymbol{q}}_\alpha} = m_\alpha \dot{\boldsymbol{q}}_\alpha \,, \tag{56}$$

$$\begin{aligned} \boldsymbol{\Pi}(\boldsymbol{r}) &= \frac{\partial L_{\text{mult}}}{\partial \dot{\boldsymbol{A}}(\boldsymbol{r})} = \frac{\partial L_{\text{rad}}}{\partial \dot{\boldsymbol{A}}(\boldsymbol{r})} - \frac{\partial}{\partial \dot{\boldsymbol{A}}(\boldsymbol{r})} \int \boldsymbol{P}^{\perp}(\boldsymbol{r}) \cdot \dot{\boldsymbol{A}}(\boldsymbol{r})\, \mathrm{d}^3 r \\ &= \varepsilon_0 \dot{\boldsymbol{A}}(\boldsymbol{r}) - \boldsymbol{P}^{\perp}(\boldsymbol{r}) = -\varepsilon_0 \boldsymbol{E}^{\perp}(\boldsymbol{r}) - \boldsymbol{P}^{\perp}(\boldsymbol{r}) \,. \end{aligned} \tag{57}$$

Since we have the relation between the electric field $\boldsymbol{E}(\boldsymbol{r})$ and the electric displacement $\boldsymbol{D}(\boldsymbol{r})$, those transverse components also satisfy

$$\boldsymbol{D}^{\perp}(\boldsymbol{r}) = \varepsilon_0 \boldsymbol{E}^{\perp}(\boldsymbol{r}) + \boldsymbol{P}^{\perp}(\boldsymbol{r}) \,, \tag{58}$$

and thus the momentum $\boldsymbol{\Pi}(\boldsymbol{r})$ can be rewritten as

$$\boldsymbol{\Pi}(\boldsymbol{r}) = -\boldsymbol{D}^{\perp}(\boldsymbol{r}) \,. \tag{59}$$

By putting them together, canonical transformation of the Lagrangian L_{mult} gives a new Hamiltonian H_{mult}

$$\begin{aligned} H_{\text{mult}} &= \sum_{\zeta,\alpha} \boldsymbol{p}_\alpha(\zeta) \cdot \dot{\boldsymbol{q}}_\alpha(\zeta) + \int \boldsymbol{\Pi}(\boldsymbol{r}) \cdot \dot{\boldsymbol{A}}(\boldsymbol{r})\, \mathrm{d}^3 r - L_{\text{mult}} \\ &= \sum_{\zeta} \left\{ \sum_{\alpha} \frac{\boldsymbol{p}_\alpha^2(\zeta)}{2m_\alpha} + V(\zeta) \right\} \\ &\quad + \left\{ \frac{1}{2} \int \left[\frac{\boldsymbol{\Pi}^2(\boldsymbol{r})}{\varepsilon_0} + \varepsilon_0 c^2 \left(\nabla \times \boldsymbol{A}(\boldsymbol{r}) \right)^2 \right] \mathrm{d}^3 r \right\} \\ &\quad + \frac{1}{\varepsilon_0} \int \boldsymbol{P}^{\perp}(\boldsymbol{r}) \cdot \boldsymbol{\Pi}(\boldsymbol{r})\, \mathrm{d}^3 r + \frac{1}{2\varepsilon_0} \int \left| \boldsymbol{P}^{\perp}(\boldsymbol{r}) \right|^2 \mathrm{d}^3 r + V_{\text{inter}} \,. \end{aligned} \tag{60}$$

It is possible to simplify (60) by separating $(1/2\varepsilon_0) \int \left| \boldsymbol{P}^{\perp}(\boldsymbol{r}) \right|^2 \mathrm{d}^3 r$ into two parts: inter- and intramolecular. Let us consider the intermolecular part:

$$\frac{1}{2\varepsilon_0} \int \boldsymbol{P}_1^{\perp}(\boldsymbol{r}) \cdot \boldsymbol{P}_2^{\perp}(\boldsymbol{r}) \mathrm{d}^3 r \,. \tag{61}$$

Noting that

$$\boldsymbol{P}_2(\boldsymbol{r}) = \boldsymbol{P}_2^{\parallel}(\boldsymbol{r}) + \boldsymbol{P}_2^{\perp}(\boldsymbol{r}) \,, \tag{62a}$$

$$\boldsymbol{P}_1^{\perp}(\boldsymbol{r}) \cdot \boldsymbol{P}_2^{\parallel}(\boldsymbol{r}) = 0 \,, \tag{62b}$$

and

$$\begin{aligned} \boldsymbol{P}_1^{\perp}(\boldsymbol{r}) \cdot \boldsymbol{P}_2^{\perp}(\boldsymbol{r}) &= \boldsymbol{P}_1^{\perp}(\boldsymbol{r}) \cdot \left\{ \boldsymbol{P}_2^{\parallel}(\boldsymbol{r}) + \boldsymbol{P}_2^{\perp}(\boldsymbol{r}) \right\} \\ &= \boldsymbol{P}_1^{\perp}(\boldsymbol{r}) \cdot \boldsymbol{P}_2(\boldsymbol{r}) \,, \end{aligned} \tag{63}$$

we rewrite (61) as follows:

$$\begin{aligned}\frac{1}{\varepsilon_0}\int \boldsymbol{P}_1^{\perp}(\boldsymbol{r}) \cdot \boldsymbol{P}_2^{\perp}(\boldsymbol{r})\,\mathrm{d}^3r &= \frac{1}{\varepsilon_0}\int \boldsymbol{P}_1^{\perp}(\boldsymbol{r}) \cdot \boldsymbol{P}_2(\boldsymbol{r})\,\mathrm{d}^3r \\ &= \sum_{i,j}\frac{\mu_i(1)\,\mu_j(2)}{\varepsilon_0}\int \delta_{ij}^{\perp}(\boldsymbol{r}-\boldsymbol{R}_1)\,\delta(\boldsymbol{r}-\boldsymbol{R}_2)\,\mathrm{d}^3r \\ &= \sum_{i,j}\frac{\mu_i(1)\,\mu_j(2)}{\varepsilon_0}\delta_{ij}^{\perp}(\boldsymbol{R}_1-\boldsymbol{R}_2) \\ &= -\sum_{i,j}\frac{\mu_i(1)\,\mu_j(2)}{4\pi\varepsilon_0 R^3}(\delta_{ij}-3\hat{e}_{Ri}\hat{e}_{Rj}) \\ &= -\frac{1}{4\pi\varepsilon_0 R^3}\left\{\boldsymbol{\mu}(1)\cdot\boldsymbol{\mu}(2)-3(\boldsymbol{\mu}(1)\cdot\boldsymbol{e}_R)(\boldsymbol{\mu}(2)\cdot\boldsymbol{e}_R)\right\}\,,\end{aligned} \tag{64}$$

where we used (51) in the first line, and the following identities for the Dirac δ function and the δ dyadics, $\delta_{ij}^{\parallel}(\boldsymbol{r})$ and $\delta_{ij}^{\perp}(\boldsymbol{r})$, in the third line:

$$\delta_{ij}\delta(\boldsymbol{r}) = \delta_{ij}^{\parallel}(\boldsymbol{r}) + \delta_{ij}^{\perp}(\boldsymbol{r})\,, \tag{65a}$$

$$\begin{aligned}\delta_{ij}^{\perp}(\boldsymbol{r}) &= -\delta_{ij}^{\parallel}(\boldsymbol{r}) \\ &= -\frac{1}{(2\pi)^3}\int \hat{e}_{ki}\hat{e}_{kj}\exp(\mathrm{i}\boldsymbol{k}\cdot\boldsymbol{r})\,\mathrm{d}^3k \\ &= \nabla_i\nabla_j\left(\frac{1}{4\pi r}\right) = -\frac{1}{4\pi r^3}(\delta_{ij}-3\hat{e}_{ri}\hat{e}_{rj})\,.\end{aligned} \tag{65b}$$

Here the subscripts i and j refer to Cartesian components, as usual. Since the exchange of the subscripts 1 and 2 gives the same result as (64), one can derive

$$\frac{1}{2\varepsilon_0}\int \boldsymbol{P}_1^{\perp}(\boldsymbol{r})\cdot\boldsymbol{P}_2^{\perp}(\boldsymbol{r})\,\mathrm{d}^3r + V_{\mathrm{inter}} = 0\,. \tag{66}$$

Therefore we can only take care of the intramolecular part of $(1/2\varepsilon_0) \times \int \left|\boldsymbol{P}^{\perp}(\boldsymbol{r})\right|^2 \mathrm{d}^3r$, and have the simplified version of H_{mult} as

$$\begin{aligned}H_{\mathrm{mult}} = &\sum_{\zeta}\left\{\sum_{\alpha}\frac{\boldsymbol{p}_\alpha^2(\zeta)}{2m_\alpha} + V(\zeta) + \frac{1}{2\varepsilon_0}\int\left|\boldsymbol{P}_\zeta^{\perp}(\boldsymbol{r})\right|^2\mathrm{d}^3r\right\} \\ &+\left\{\frac{1}{2}\int\left[\frac{\boldsymbol{\Pi}^2(\boldsymbol{r})}{\varepsilon_0} + \varepsilon_0 c^2(\nabla\times\boldsymbol{A}(\boldsymbol{r}))^2\right]\mathrm{d}^3r\right\} \\ &+\frac{1}{\varepsilon_0}\int \boldsymbol{P}^{\perp}(\boldsymbol{r})\cdot\boldsymbol{\Pi}(\boldsymbol{r})\,\mathrm{d}^3r\,,\end{aligned} \tag{67}$$

where each line represents the charged particle motion in each molecule, free electromagnetic field, and the interaction, respectively. Since we can expand the polarization density in terms of 2^ℓ multipoles ($\ell = 1, 2, 3, \cdots$), as shown in (50), we call H_{mult} the multipole Hamiltonian. The interaction part can be more explicitly written as

$$\begin{aligned}
\frac{1}{\varepsilon_0}\int \boldsymbol{P}^{\perp}(\boldsymbol{r})\cdot\boldsymbol{\Pi}(\boldsymbol{r})\,\mathrm{d}^3r &= -\frac{1}{\varepsilon_0}\int \boldsymbol{P}^{\perp}(\boldsymbol{r})\cdot\boldsymbol{D}^{\perp}(\boldsymbol{r})\,\mathrm{d}^3r \\
&= -\frac{1}{\varepsilon_0}\int \boldsymbol{P}(\boldsymbol{r})\cdot\boldsymbol{D}^{\perp}(\boldsymbol{r})\,\mathrm{d}^3r \\
&= -\frac{1}{\varepsilon_0}\left\{\boldsymbol{\mu}(1)\cdot\boldsymbol{D}^{\perp}(\boldsymbol{R}_1)+\boldsymbol{\mu}(2)\cdot\boldsymbol{D}^{\perp}(\boldsymbol{R}_2)\right\}
\end{aligned} \tag{68}$$

with the help of (51) and (59). When the considered system is quantized, quantities such as $\boldsymbol{\mu}(i)$ and $\boldsymbol{D}^{\perp}(\boldsymbol{R}_i)$ $(i = 1, 2)$ should be replaced by the corresponding operators,

$$-\frac{1}{\varepsilon_0}\left\{\hat{\boldsymbol{\mu}}(1)\cdot\hat{\boldsymbol{D}}^{\perp}(\boldsymbol{R}_1)+\hat{\boldsymbol{\mu}}(2)\cdot\hat{\boldsymbol{D}}^{\perp}(\boldsymbol{R}_2)\right\}, \tag{69}$$

yielding the quantized multipole Hamiltonian.

5 Elementary Excitation Modes and Electronic Polarization

The concept of elementary excitations, or quasiparticles, has been discussed for a long time, and it is valuable for description of excited states, complex behavior, or dynamics of a many-body system [27–32]. Excited states of a many-body system are considered a collection of certain fundamental excited states that we call the elementary excitation. As a prerequisite, there must be a well-defined excitation energy whose value should be larger than the width of the relevant energy level. Then the relation between momentum $\boldsymbol{p}$ and energy E of the elementary excitation, i.e., $E = E(\boldsymbol{p})$, is referred to as the dispersion relation.

Phonons, as quanta of normal modes of crystal vibration, are well known as an example of the elementary excitation modes in a solid. The motion is collective, which means that the total number of Phonons is independent of the number of crystal lattice. The momentum of the elementary excitation is $\boldsymbol{p} = \hbar\boldsymbol{k}$ in terms of the wave vector $\boldsymbol{k}$ of normal vibration, not the mechanical momentum of crystal lattice itself. The energy of the elementary excitation is also given by the angular frequency ω of the normal vibration as $E = \hbar\omega$.

As other examples of the elementary excitations, we have plasmons, which correspond to collective motion of electron density in interacting electron gas; polarons are quasiparticles originated from the coupling between conduction electrons and optical Phonons; and magnons, corresponding to collective modes of spin-density waves. Excitons are also well known, and describe the elementary excitation related to an electron–hole pair in a solid. As a limiting case, Frenkel excitons and Wannier excitons are frequently discussed.

When the distance between the electron and hole in an exciton (Bohr radius of the exciton) is smaller than the atomic distance in the crystal, it is called a Frenkel exciton; Wannier excitons correspond to the opposite case, in which the Bohr radius of the exciton is larger than the lattice constant of the crystal.

5.1 Polaritons and Electronic Polarization

Let us consider the light–matter interaction on the basis of the exciton concept. Incident photons interact with matter, and produce the successive creation and annihilation of excitons and photons in matter, i.e., an electronic polarization field. This process indicates a new steady state with a new dispersion relation and energy due to the photon–exciton interaction. Normal modes, or elementary excitation modes for this coupled oscillation, are called polaritons. In particular, they are called exciton polaritons due to the occurrence of the mixed states of photons and excitons. The situation is analogous to the case in which two coupled oscillations with angular frequencies ω_1 and ω_2 produce new normal oscillations with angular frequencies Ω_1 and Ω_2. Dressed atom states in an atom–photon interacting system [33] are conceptually similar to the normal modes of the photon and electronic polarization field, or exciton polaritons as quasiparticles.

Rewriting the Hamiltonian for a photon–electron interacting system in terms of excitons, one can obtain the following Hamiltonian describing exciton polaritons:

$$\hat{H} = \sum_{\boldsymbol{k}} \hbar\omega_{\boldsymbol{k}} \hat{a}_{\boldsymbol{k}}^{\dagger} \hat{a}_{\boldsymbol{k}} + \sum_{\boldsymbol{k}} \hbar\varepsilon_{\boldsymbol{k}} \hat{b}_{\boldsymbol{k}}^{\dagger} \hat{b}_{\boldsymbol{k}} + \sum_{\boldsymbol{k}} \hbar D \left(\hat{a}_{\boldsymbol{k}} + \hat{a}_{-\boldsymbol{k}}^{\dagger}\right) \left(\hat{b}_{\boldsymbol{k}}^{\dagger} + \hat{b}_{-\boldsymbol{k}}\right) . \tag{70}$$

Here the first and second terms correspond to the Hamiltonians for free photons and free excitons, respectively, and the third term describes the photon–exciton interaction, whose coupling strength is $\hbar D$. The explicit expression for $\hbar D$ will be given in Sect. 6.3. Energies due to zero-point oscillation are suppressed in (70). Creation and annihilation operators for photons are denoted by $\hat{a}_{\boldsymbol{k}}^{\dagger}$ and $\hat{a}_{\boldsymbol{k}}$, while those for excitons are designated $\hat{b}_{\boldsymbol{k}}^{\dagger}$ and $\hat{b}_{\boldsymbol{k}}$. In the rewriting process, we define the creation and annihilation operators of excitons to be

$$\hat{b}_{\boldsymbol{l}}^{\dagger} = \hat{c}_{\boldsymbol{l},\mathrm{c}}^{\dagger} \hat{c}_{\boldsymbol{l},\mathrm{v}} , \tag{71a}$$

$$\hat{b}_{\boldsymbol{l}} = \hat{c}_{\boldsymbol{l},\mathrm{v}}^{\dagger} \hat{c}_{\boldsymbol{l},\mathrm{c}} , \tag{71b}$$

where we use the operator $\hat{c}_{\boldsymbol{l},\mathrm{v}}$ that annihilates an electron in the valence band v within an atom at the lattice site $\boldsymbol{l}$, and its Hermitian conjugate operator $\hat{c}_{\boldsymbol{l},\mathrm{v}}^{\dagger}$, as well as the operator $\hat{c}_{\boldsymbol{l},\mathrm{c}}^{\dagger}$ that creates an electron in the conduction band c within an atom at the lattice site $\boldsymbol{l}$, and its Hermitian conjugate operator $\hat{c}_{\boldsymbol{l},c}$. According to the conventional method, in addition, we introduce the operators

$$\hat{b}_{\boldsymbol{k}} = \frac{1}{\sqrt{N}} \sum_{\boldsymbol{l}} \mathrm{e}^{-\mathrm{i}\boldsymbol{k}\cdot\boldsymbol{l}} \hat{b}_{\boldsymbol{l}} , \tag{72a}$$

$$\hat{b}_{\boldsymbol{k}}^{\dagger} = \frac{1}{\sqrt{N}} \sum_{\boldsymbol{l}} \mathrm{e}^{\mathrm{i}\boldsymbol{k}\cdot\boldsymbol{l}} \hat{b}_{\boldsymbol{l}}^{\dagger} , \tag{72b}$$

in the momentum representation. Here the total number of lattice sites is assumed to be N in the crystal considered.

Once the Hamiltonian for exciton polaritons is given by (70), one can obtain eigenstates and eigenenergies of exciton polaritons C, or the dispersion relation. For simplicity, we adopt the rotating wave approximation and neglect terms $\hat{a}_{-\boldsymbol{k}}^{\dagger}\hat{b}_{\boldsymbol{k}}^{\dagger}$ and $\hat{a}_{\boldsymbol{k}}\hat{b}_{-\boldsymbol{k}}$ which create or annihilate a photon and exciton at the same time, and consider the Hamiltonian

$$\hat{H} = \sum_{\boldsymbol{k}} \hat{H}_{\boldsymbol{k}} , \tag{73a}$$

$$\hat{H}_{\boldsymbol{k}} = \hbar \left(\omega_{\boldsymbol{k}} \hat{a}_{\boldsymbol{k}}^{\dagger} \hat{a}_{\boldsymbol{k}} + \varepsilon_{\boldsymbol{k}} \hat{b}_{\boldsymbol{k}}^{\dagger} \hat{b}_{\boldsymbol{k}} \right) + \hbar D \left(\hat{b}_{\boldsymbol{k}}^{\dagger} \hat{a}_{\boldsymbol{k}} + \hat{a}_{\boldsymbol{k}}^{\dagger} \hat{b}_{\boldsymbol{k}} \right) . \tag{73b}$$

We next introduce the creation and annihilation operators of exciton polaritons, $\hat{\xi}_1^{\dagger}$, $\hat{\xi}_2^{\dagger}$ and $\hat{\xi}_1$, $\hat{\xi}_2$, corresponding to new eigenfrequencies $\Omega_{\boldsymbol{k},1}$, $\Omega_{\boldsymbol{k},2}$, respectively. The Hamiltonian $\hat{H}_{\boldsymbol{k}}$ is assumed to be diagonalized in terms of $\hat{\xi}_1$, $\hat{\xi}_2$, and we rewrite (73b) as

$$\begin{aligned}
\hat{H}_{\boldsymbol{k}} &= \hbar \left(\Omega_{\boldsymbol{k},1} \hat{\xi}_1^{\dagger} \hat{\xi}_1 + \Omega_{\boldsymbol{k},2} \hat{\xi}_2^{\dagger} \hat{\xi}_2 \right) \\
&= \hbar \left(\hat{b}_{\boldsymbol{k}}^{\dagger}, \hat{a}_{\boldsymbol{k}}^{\dagger} \right) A \begin{pmatrix} \hat{b}_{\boldsymbol{k}} \\ \hat{a}_{\boldsymbol{k}} \end{pmatrix} \\
&= \hbar \left(a_{11} \hat{b}_{\boldsymbol{k}}^{\dagger} \hat{b}_{\boldsymbol{k}} + a_{12} \hat{b}_{\boldsymbol{k}}^{\dagger} \hat{a}_{\boldsymbol{k}} + a_{21} \hat{a}_{\boldsymbol{k}}^{\dagger} \hat{b}_{\boldsymbol{k}} + a_{22} \hat{a}_{\boldsymbol{k}}^{\dagger} \hat{a}_{\boldsymbol{k}} \right) ,
\end{aligned} \tag{74}$$

where A is the 2 by 2 matrix whose elements are given by

$$A = \begin{bmatrix} a_{11} & a_{12} \\ a_{21} & a_{22} \end{bmatrix} = \begin{bmatrix} \varepsilon_{\boldsymbol{k}} & D \\ D & \omega_{\boldsymbol{k}} \end{bmatrix} . \tag{75}$$

Applying the unitary transformation U, (i.e., $U^{\dagger} = U^{-1}$)

$$\begin{pmatrix} \hat{b}_{\boldsymbol{k}} \\ \hat{a}_{\boldsymbol{k}} \end{pmatrix} = U \begin{pmatrix} \hat{\xi}_1 \\ \hat{\xi}_2 \end{pmatrix} \quad \text{with} \quad U = \begin{bmatrix} u_{11} & u_{12} \\ u_{21} & u_{22} \end{bmatrix} \tag{76}$$

to (74), we have

$$\hbar \left(\hat{b}_{\boldsymbol{k}}^{\dagger}, \hat{a}_{\boldsymbol{k}}^{\dagger} \right) A \begin{pmatrix} \hat{b}_{\boldsymbol{k}} \\ \hat{a}_{\boldsymbol{k}} \end{pmatrix} = \hbar \left(\hat{\xi}_1^{\dagger}, \hat{\xi}_2^{\dagger} \right) U^{\dagger} A U \begin{pmatrix} \hat{\xi}_1 \\ \hat{\xi}_2 \end{pmatrix} . \tag{77}$$

Since $U^\dagger AU = U^{-1}AU$ is diagonal, we put

$$U^{-1}AU = \begin{bmatrix} \Omega_{\boldsymbol{k},1} & 0 \\ 0 & \Omega_{\boldsymbol{k},2} \end{bmatrix} \equiv \Lambda \,, \tag{78}$$

and obtain $AU = U\Lambda$, which reduces in terms of components ($j = 1, 2$) to

$$\begin{bmatrix} \varepsilon_{\boldsymbol{k}} - \Omega_{\boldsymbol{k},j} & D \\ D & \omega_{\boldsymbol{k}} - \Omega_{\boldsymbol{k},j} \end{bmatrix} \begin{pmatrix} u_{1j} \\ u_{2j} \end{pmatrix} = 0 \,. \tag{79}$$

This immediately gives the eigenvalue equation

$$(\Omega - \varepsilon_{\boldsymbol{k}})(\Omega - \omega_{\boldsymbol{k}}) - D^2 = 0 \,, \tag{80}$$

and the eigenenergies of exciton polaritons are

$$\hbar\Omega_{\boldsymbol{k},j} = \hbar \left[\frac{\varepsilon_{\boldsymbol{k}} + \omega_{\boldsymbol{k}}}{2} \pm \frac{\sqrt{(\varepsilon_{\boldsymbol{k}} - \omega_{\boldsymbol{k}})^2 + 4D^2}}{2} \right] . \tag{81}$$

Equation (81) provides the new dispersion relation that we are looking for. Using the dispersion relation of photons $\omega_{\boldsymbol{k}} = ck$ with $k = |\boldsymbol{k}|$, we can plot eigenenergies of exciton polaritons as a function of k, as shown in Fig. 1. Here, for simplicity, we approximate the exciton dispersion as $\varepsilon_{\boldsymbol{k}} = \hbar\Omega$, independent of k.

From (79) and the unitarity of U, we have for the components of the eigenvectors

$$u_{2j} = -\frac{\varepsilon_{\boldsymbol{k}} - \Omega_{\boldsymbol{k},j}}{D} u_{1j} \,, \tag{82a}$$

$$u_{1j}^2 + u_{2j}^2 = 1 \,, \tag{82b}$$

and thus

$$\left\{ 1 + \left(\frac{\varepsilon_{\boldsymbol{k}} - \Omega_{\boldsymbol{k},j}}{D} \right)^2 \right\} u_{1j}^2 = 1 \,. \tag{83}$$

Finally, the eigenvectors of exciton polaritons are

$$u_{1j} = \left\{ 1 + \left(\frac{\varepsilon_{\boldsymbol{k}} - \Omega_{\boldsymbol{k},j}}{D} \right)^2 \right\}^{-1/2} , \tag{84a}$$

$$u_{2j} = -\left(\frac{\varepsilon_{\boldsymbol{k}} - \Omega_{\boldsymbol{k},j}}{D} \right) \left\{ 1 + \left(\frac{\varepsilon_{\boldsymbol{k}} - \Omega_{\boldsymbol{k},j}}{D} \right)^2 \right\}^{-1/2} . \tag{84b}$$

New steady states for exciton polaritons can be described by (81), (84a), and (84b).

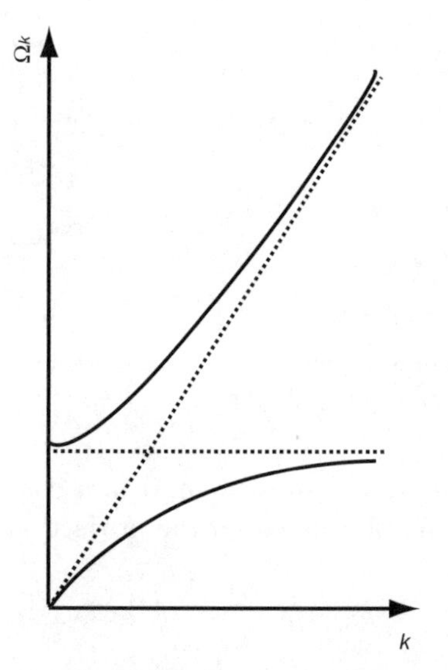

Fig. 1. Dispersion relation of exciton polaritons. Energies of exciton polaritons Ω_k shown as solid curves are schematically plotted as a function of k

6 Optical Near-Field Interaction: Yukawa Potential

Based on the concepts and methods introduced in the preceding sections, we investigate the formulation of an optical near-field system that was briefly mentioned in Sect. 1. Moreover, explicit functional forms of the optical near-field interaction will be obtained by using either the effective interaction $\hat{V}_{\text{eff}}$ in (26), or its approximation given in (37) [18,19].

6.1 Relevant Microscopic Subsystem and Irrelevant Macroscopic Subsystem

As mentioned in Sect. 1, the optical near-field system consists of two subsystems: one is a macroscopic subsystem whose typical dimension is much larger than the wavelength of the incident light. The other is a nanometric subsystem whose constituents are a nanometric aperture or a protrusion at the apex of the near-field optical probe, and a nanometric sample. We call such an aperture or protrusion a probe tip. As a nanometric sample we mainly take a single atom/molecule or quantum dot (QD). Subdivision of the total system is schematically illustrated in Fig. 2. The two subsystems are interacting with each other, and it is very important to formulate the interactions consistently and systematically.

We call the nanometric subsystem the relevant subsystem n, and the macroscopic subsystem the irrelevant subsystem M. We are interested in the subsystem n, in particular the interaction induced in the subsystem n. There-

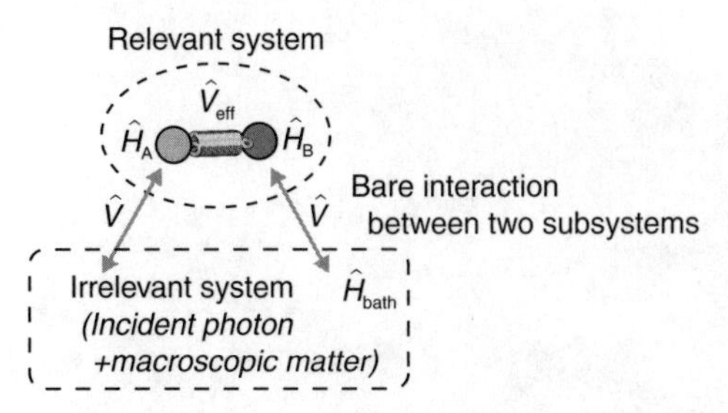

Fig. 2. Subdivision of an optical near-field system is schematically illustrated. One is the relevant subsystem, and the other is the irrelevant subsystem

fore it is important to renormalize effects due to the subsystem M in a consistent and systematic way. We show a formulation based on the projection operator method described in Sects. 2 and 3.

6.2 *P* Space and *Q* Space

It is preferable to express exact states $|\psi\rangle$ for the total system in terms of a small number of bases of as a few degrees of freedom as possible that span P space. In the following we assume two states as the P-space components: $|\phi_1\rangle = |s^*\rangle\,|p\rangle \otimes \left|0_{(M)}\right\rangle$ and $|\phi_2\rangle = |s\rangle\,|p^*\rangle \otimes \left|0_{(M)}\right\rangle$. Here $|s\rangle$ and $|s^*\rangle$ are eigenstates of the sample, which is isolated, while $|p\rangle$ and $|p^*\rangle$ are eigenstates of the probe tip, which is also isolated. Resonant excitation energies of the sample and the probe tip are assumed to be $E_s = \hbar\Omega_0\,(s)$ and $E_p = \hbar\Omega_0\,(p)$, respectively. In addition, exciton–polariton states as bases discussed in Sect. 5 are used to describe the macroscopic subsystem M, and thus $\left|0_{(M)}\right\rangle$ represents the vacuum for exciton polaritons. Note that there exist photons and electronic matter excitations even in the vacuum state $\left|0_{(M)}\right\rangle$. The direct product is denoted by the symbol $\otimes$. The complementary space to the P space is called Q space, which is spanned by a huge number of bases of a large number of degrees of freedom not included in the P space.

6.3 Effective Interaction in the Nanometric Subsystem

When we evaluate the effective interaction given by (37) in the P space discussed above and trace out the other degrees of freedom, the result gives the effective interaction potential of the nanometric subsystem n after renormalizing the effects from the macroscopic subsystem M. Using the effective interaction potential, one can ignore the subsystem M, as if the subsystem n were isolated and separated from the subsystem M.

As the first step of the procedure, we employ the "bare interaction" between the two subsystems:

$$\hat{V} = -\frac{1}{\varepsilon_0}\left\{\hat{\boldsymbol{\mu}}_s \cdot \hat{\boldsymbol{D}}^{\perp}(\boldsymbol{r}_s) + \hat{\boldsymbol{\mu}}_p \cdot \hat{\boldsymbol{D}}^{\perp}(\boldsymbol{r}_p)\right\} \tag{85}$$

(see (69)). It should be noted that there are no interactions without incident photons in the macroscopic subsystem M. The subscripts s and p represent physical quantities related to the sample and the probe tip, respectively. Representative positions of the sample and the probe tip are chosen for simplicity to be $\boldsymbol{r}_s$ and $\boldsymbol{r}_p$, respectively, but those may be composed of several positions. In that case the quantities in curly brackets in (85) should be summed over. The transverse component of the electric displacement operator, $\hat{\boldsymbol{D}}^{\perp}(\boldsymbol{r})$, can be expressed in terms of the vector potential $\hat{\boldsymbol{A}}(\boldsymbol{r})$ and its conjugate momentum $\hat{\boldsymbol{\Pi}}(\boldsymbol{r})$ from the multipole Lagrangian $\hat{L}_{\mathrm{mult}}$:

$$\begin{aligned}\hat{\boldsymbol{\Pi}}(\boldsymbol{r}) &= \frac{\partial \hat{L}_{\mathrm{mult}}}{\partial \dot{\hat{\boldsymbol{A}}}} = \varepsilon_0 \frac{\partial \hat{\boldsymbol{A}}}{\partial t} - \hat{\boldsymbol{P}}^{\perp}(\boldsymbol{r}) \\ &= -\varepsilon_0 \hat{\boldsymbol{E}}^{\perp}(\boldsymbol{r}) - \hat{\boldsymbol{P}}^{\perp}(\boldsymbol{r}) = -\hat{\boldsymbol{D}}^{\perp}(\boldsymbol{r}) \ . \end{aligned} \tag{86}$$

With the help of the mode expansion of $\hat{\boldsymbol{A}}(\boldsymbol{r})$

$$\hat{\boldsymbol{A}}(\boldsymbol{r}) = \sum_{\boldsymbol{k}} \sum_{\lambda=1}^{2} \left(\frac{\hbar}{2\varepsilon_0 V \omega_{\boldsymbol{k}}} \right)^{1/2} \boldsymbol{e}_\lambda(\boldsymbol{k}) \left\{ \hat{a}_\lambda(\boldsymbol{k}) \mathrm{e}^{\mathrm{i}\boldsymbol{k}\cdot\boldsymbol{r}} + \hat{a}_\lambda^\dagger(\boldsymbol{k}) \mathrm{e}^{-\mathrm{i}\boldsymbol{k}\cdot\boldsymbol{r}} \right\} \tag{87}$$

and $\hat{\boldsymbol{\Pi}}(\boldsymbol{r})$ from the minimal coupling Lagrangian $\hat{L}$

$$\begin{aligned}\hat{\boldsymbol{\Pi}}(\boldsymbol{r}) &= \frac{\partial \hat{L}}{\partial \dot{\hat{\boldsymbol{A}}}} = \varepsilon_0 \frac{\partial \hat{\boldsymbol{A}}}{\partial t} \\ &= -\mathrm{i} \sum_{\boldsymbol{k}} \sum_{\lambda=1}^{2} \left(\frac{\varepsilon_0 \hbar \omega_{\boldsymbol{k}}}{2V} \right)^{1/2} \boldsymbol{e}_\lambda(\boldsymbol{k}) \left\{ \hat{a}_\lambda(\boldsymbol{k}) \mathrm{e}^{\mathrm{i}\boldsymbol{k}\cdot\boldsymbol{r}} - \hat{a}_\lambda^\dagger(\boldsymbol{k}) \mathrm{e}^{-\mathrm{i}\boldsymbol{k}\cdot\boldsymbol{r}} \right\} \ , \end{aligned} \tag{88}$$

we can rewrite (86) as

$$\hat{\boldsymbol{D}}^{\perp}(\boldsymbol{r}) = \mathrm{i} \sum_{\boldsymbol{k}} \sum_{\lambda=1}^{2} \left(\frac{\varepsilon_0 \hbar \omega_{\boldsymbol{k}}}{2V} \right)^{1/2} \boldsymbol{e}_\lambda(\boldsymbol{k}) \left\{ \hat{a}_\lambda(\boldsymbol{k}) \mathrm{e}^{\mathrm{i}\boldsymbol{k}\cdot\boldsymbol{r}} - \hat{a}_\lambda^\dagger(\boldsymbol{k}) \mathrm{e}^{-\mathrm{i}\boldsymbol{k}\cdot\boldsymbol{r}} \right\} \ , \tag{89}$$

where the creation and annihilation operators of a photon with wave vector $\boldsymbol{k}$, angular frequency $\omega_{\boldsymbol{k}}$, and polarization component λ are designated as $\hat{a}_\lambda^\dagger(\boldsymbol{k})$ and $\hat{a}_\lambda(\boldsymbol{k})$, respectively. The quantization volume is V, and the unit vector related to the polarization direction is $\boldsymbol{e}_\lambda(\boldsymbol{k})$.

Since exciton–polariton states are employed as bases to describe the macroscopic subsystem M, the creation and annihilation operators of a photon in (89) can be rewritten in terms of the exciton–polariton operators $\hat{\xi}^\dagger(\boldsymbol{k})$ and $\hat{\xi}(\boldsymbol{k})$, and then substituted into (85). Using the electric dipole operator

$$\hat{\boldsymbol{\mu}}_\alpha = \left(\hat{B}(\boldsymbol{r}_\alpha) + \hat{B}^\dagger(\boldsymbol{r}_\alpha) \right) \boldsymbol{\mu}_\alpha \ , \tag{90}$$

we obtain the "bare interaction" in the exciton–polariton picture:

$$\hat{V} = -\mathrm{i}\left(\frac{\hbar}{2\varepsilon_0 V}\right)^{1/2} \sum_{\alpha=s}^{p} \sum_{\boldsymbol{k}} \left(\hat{B}(\boldsymbol{r}_\alpha) + \hat{B}^\dagger(\boldsymbol{r}_\alpha)\right) \times \left(K_\alpha(\boldsymbol{k})\hat{\xi}(\boldsymbol{k}) - K_\alpha^*(\boldsymbol{k})\hat{\xi}^\dagger(\boldsymbol{k})\right). \tag{91}$$

Here $K_\alpha(\boldsymbol{k})$ is the coupling coefficient between the exciton polariton and the nanometric subsystem n, and is given by

$$K_\alpha(\boldsymbol{k}) = \sum_{\lambda=1}^{2} (\boldsymbol{\mu}_\alpha \cdot \boldsymbol{e}_\lambda(\boldsymbol{k})) f(k)\, \mathrm{e}^{\mathrm{i}\boldsymbol{k}\cdot\boldsymbol{r}_\alpha} \tag{92}$$

with

$$f(k) = \frac{ck}{\sqrt{\Omega(k)}} \sqrt{\frac{\Omega^2(k) - \Omega^2}{2\Omega^2(k) - \Omega^2 - (ck)^2}}\,. \tag{93}$$

The asterisk in (91) stands for the complex conjugate, while c, $\Omega(k)$, and Ω in (93) are the velocity of light in vacuum and the eigenfrequencies of the exciton polariton and electronic polarization of the macroscopic subsystem M, respectively [34]. Note that the wave-number dependence of $f(k)$ characterizes a typical interaction range of exciton polaritons coupled to the nanometric subsystem n.

The next step is to evaluate the amplitude of effective sample–probe tip interaction in the P space:

$$V_{\mathrm{eff}}(2,1) \equiv \langle\phi_2|\,\hat{V}_{\mathrm{eff}}\,|\phi_1\rangle\,. \tag{94}$$

Using (37) as $\hat{V}_{\mathrm{eff}}$, we can explicitly write (94) in the form

$$\begin{aligned} V_{\mathrm{eff}}(2,1) &= 2\,\langle\phi_2|\,P\hat{V}Q\left(E_P^0 - E_Q^0\right)^{-1} Q\hat{V}P\,|\phi_1\rangle \\ &= 2\sum_m \langle\phi_2|\,P\hat{V}Q\,|m\rangle\,\langle m|Q\left(E_P^0 - E_Q^0\right)^{-1} Q\hat{V}P\,|\phi_1\rangle\,, \end{aligned} \tag{95}$$

where the second line shows that a virtual transition from the initial state $|\phi_1\rangle$ in the P space to an intermediate state $|m\rangle$ in the Q space is followed by a subsequent virtual transition from the intermediate state $|m\rangle$ to the final state $|\phi_2\rangle$ in the P space. We can then proceed by substituting the explicit "bare interaction" $\hat{V}$ in (91) with (92) and (93) into (95). Since one exciton–polariton state, as for the subsystem M within an arbitrary intermediate state $|m\rangle$, can only contribute to nonzero matrix elements, (95) can be transformed into

$$V_{\mathrm{eff}}(2,1) = -\frac{1}{(2\pi)^3\varepsilon_0} \int \mathrm{d}^3k \left[\frac{K_p(\boldsymbol{k})K_s^*(\boldsymbol{k})}{\Omega(k) - \Omega_0(s)} + \frac{K_s(\boldsymbol{k})K_p^*(\boldsymbol{k})}{\Omega(k) + \Omega_0(p)}\right], \tag{96}$$

where summation over $\boldsymbol{k}$ is replaced by integration over $\boldsymbol{k}$ in the usual manner.

Interchanging the arguments 1 and 2 (the role of the sample and probe tip), we can similarly calculate $V_{\text{eff}}(1,2) \equiv \langle\phi_1|\hat{V}_{\text{eff}}|\phi_2\rangle$:

$$V_{\text{eff}}(1,2) = -\frac{1}{(2\pi)^3\varepsilon_0}\int \mathrm{d}^3k \left[\frac{K_s(\boldsymbol{k})K_p^*(\boldsymbol{k})}{\Omega(k)-\Omega_0(p)} + \frac{K_p(\boldsymbol{k})K_s^*(\boldsymbol{k})}{\Omega(k)+\Omega_0(s)}\right]. \quad (97)$$

Therefore the total amplitude of the effective sample–probe tip interaction is given by the sum of (96) and (97), which includes effects from the macroscopic subsystem M. We write this effective interaction potential for the nanometric subsystem n as $V_{\text{eff}}(r)$ because we have both $V_{\text{eff}}(2,1)$ and $V_{\text{eff}}(1,2)$ as a function of the distance between the sample and the probe tip after the $\boldsymbol{k}$-integration.

6.4 Effective Mass Approximation of Exciton Polaritons and Yukawa Potential

The dispersion relation of exciton polaritons, as schematically shown in Fig. 1, can be approximated by

$$\hbar\Omega(\boldsymbol{k}) = \hbar\Omega + \frac{(\hbar\boldsymbol{k})^2}{2m_{\text{pol}}} \quad (98)$$

in terms of the effective mass of exciton polaritons m_{pol}. From this and the discussion of (95), it follows that $V_{\text{eff}}(r)$ results from the exchange of virtual exciton polaritons, or massive virtual photons. Substituting (98) into (96) and (97), we can rewrite the effective potential $V_{\text{eff}}(r)$ as

$$\begin{aligned}
V_{\text{eff}}(r) = &-\frac{4\mu_{\text{s}}\mu_{\text{p}}\hbar E_{\text{pol}}}{3\mathrm{i}\pi r(\hbar c)^2}\int \mathrm{d}k\, k f^2(k)\,\mathrm{e}^{\mathrm{i}kr} \\
&\left\{\frac{1}{k^2+2E_{\text{pol}}(E_{\text{m}}+E_s)(\hbar c)^{-2}} + \frac{1}{k^2+2E_{\text{pol}}(E_{\text{m}}-E_s)(\hbar c)^{-2}}\right. \\
&\left.+\frac{1}{k^2+2E_{\text{pol}}(E_{\text{m}}+E_p)(\hbar c)^{-2}} + \frac{1}{k^2+2E_{\text{pol}}(E_{\text{m}}-E_p)(\hbar c)^{-2}}\right\} \\
= &\frac{2\mu_{\text{s}}\mu_{\text{p}}E_{\text{pol}}^2}{3\mathrm{i}\pi r(\hbar c)^2}\int \mathrm{d}k\, kF(k)\,e^{\mathrm{i}kr}\,, \quad (99)
\end{aligned}$$

where we use the notation $E_{\text{pol}} = m_{\text{pol}}c^2$ and $E_{\text{m}} = \hbar\Omega$, and average the summation over λ as 2/3. In addition, we define $F(k)$ as

$$\begin{aligned}
F(k) \equiv &\left(\frac{A_+}{k^2+\Delta_{s+}^2} - \frac{A_-}{k^2+\Delta_{s-}^2}\right) + \left(\frac{B_+}{k^2+\Delta_{p+}^2} - \frac{B_-}{k^2+\Delta_{p-}^2}\right) \\
&+\left(\frac{C_+}{k^2+\Delta_{m+}^2} - \frac{C_-}{k^2+\Delta_{m-}^2}\right), \quad (100)
\end{aligned}$$

with

$$\Delta_{s\pm} = \frac{\sqrt{2E_{\rm pol}\left(E_{\rm m} \pm E_s\right)}}{\hbar c}, \tag{101a}$$

$$\Delta_{p\pm} = \frac{\sqrt{2E_{\rm pol}\left(E_{\rm m} \pm E_p\right)}}{\hbar c}, \tag{101b}$$

which is related to the four kinds of effective masses of the Yukawa potential, as will soon be seen. The effective masses with the suffix + are larger than those with the suffix −, and thus Δ_{s+} (Δ_{p+}) is heavier and has shorter interaction range than Δ_{s-} (Δ_{p-}). The two constants Δ_{m+} and Δ_{m-} mainly give the periodic functions related to the property of the macroscopic matter subsystem, not to the microscopic subsystem, and the third pair of terms in (100) will be omitted in the following. The weight factors in (100), that is $A_\pm$ and $B_\pm$, are functions of E_s, E_p , $E_{\rm pol}$, and E_m, and are given by

$$A_\pm = \pm\frac{2\left(E_m \mp E_s\right)\left(E_m \pm E_s\right)^2}{E_s\left(E_s \pm W_+ E_m\right)\left(E_s \pm W_- E_m\right)}, \tag{102a}$$

$$B_\pm = \pm\frac{2\left(E_m \mp E_p\right)\left(E_m \pm E_p\right)^2}{E_p\left(E_p \pm W_+ E_m\right)\left(E_p \pm W_- E_m\right)}, \tag{102b}$$

with

$$W_\pm = \frac{E_{\rm pol}}{2E_m} \pm \sqrt{\left\{1 - \frac{E_{\rm pol}}{2E_m}\right\}^2 - \frac{1}{2}}. \tag{103}$$

Finally, as the sum of the Yukawa functions with several kinds of masses, we obtain the following effective potential, or near-field optical potential $V_{\rm eff}(r)$:

$$\begin{aligned} V_{\rm eff}(r) = & \frac{2\mu_s\mu_p E_{\rm pol}^2}{3\left(\hbar c\right)^2}\left\{A_+ Y\left(\Delta_{s+} r\right) - A_- Y\left(\Delta_{s-} r\right)\right. \\ & \left. + B_+ Y\left(\Delta_{p+} r\right) - B_- Y\left(\Delta_{p-} r\right)\right\}, \end{aligned} \tag{104a}$$

$$Y(\kappa r) \equiv \frac{\exp\left(-\kappa r\right)}{r}. \tag{104b}$$

To sum up, we find that the major part of the effective interaction exerted in the microscopic subsystem n is the Yukawa potential, after renormalizing the effects from the macroscopic subsystem. This interaction is mediated by massive virtual photons, or polaritons, where exciton polaritons have been employed in the explicit formulation, but in principle other types of polaritons should be applicable.

In this section we have mainly focused on the effective interaction of the microscopic subsystem n, after tracing out the other degrees of freedoms. It is certainly possible to have a formulation with the projection onto the P space that is spanned in terms of the degrees of freedoms of the massive virtual photons. This kind of formulation emphasizes a "dressed photon" picture, in which photons are not massless but massive due to the matter excitations.

7 Applications

Let us now turn to some applications of our formulation. Although there might be a variety of possibilities, we address basic problems related to atom deflection, trapping, and manipulation with optical near fields generated in a probe tip–atom system, which is expected to be developed towards nano photonics and atom photonics. At the same time we also apply our approach to some conventional optical near-field problems that have been discussed from classical electrodynamics viewpoints.

7.1 Single Atom Manipulation

Previous Studies

In 1990 D. Eigler et al. manipulated both Xe atoms on a Ni substrate and CO molecules on a Pt substrate using a scanning tunneling microscope (STM) to demonstrate atom tweezers and nano-fabrication [35–37]. A single Cs atom was trapped in a cavity with a length of a few tens micrometers by H.J. Kimble et al., who utilized propagating light and laser cooling techniques that made rapid progress during these years [38]. Since the optical near field is independent of the diffraction limit that the propagating light suffers from, one can expect more precise control of an atom with the optical near field than that with propagating light. H. Hori and M. Ohtsu showed that a variety of atoms could be trapped near a nanometric near-field optical probe tip if the optical frequency of the incident light is negatively detuned from an atomic resonant frequency, i.e., red detuned [8,39,40]. V.V. Klimov et al. estimated the spatial distribution of optical near fields required for an atom trap [41]. H. Ito et al. proposed a method to trap a single Rb atom with an optical near field, where they assumed a potential well due to the balance of attractive van der Waals forces and repulsive dipole forces exerted in a near-field optical probe tip and the atom system [42,43]. They phenomenologically obtained the potential well, assuming the spatial distribution of the optical near field described by the Yukawa function. This assumption is now being verified by an experiment with two nanometric probe tips [43].

These studies are very important and intriguing from the following viewpoints: optical near field observation with an atom probe, nano-structure fabrication with single atom manipulation, and local state-control of nanostructure, etc. We will thus devote ourselves to these themes in the rest of the section, on the basis of the formulation that we have developed.

Possibility of a Single-Atom Trap with Optical Near Fields

Let us first take an example with a typical alkali-metal atom ^{85}Rb, in order to qualitatively examine the features of the effective potential described by (104a). We take $E_s = 1.59$ eV for the excitation energy between the $5S_{1/2}$ and

$5P_{3/2}$ levels of ^{85}Rb. Assuming infrared and visible excitations of a macroscopic matter system and a probe tip, we vary the values of E_m and E_p over the range $1.0 \leq E_m \leq 1.8$ eV and $1.0 \leq E_p \leq 1.2$ eV, respectively.

Then how does the near-field optical potential $V_{\text{eff}}(r)$ change? Figure 3a–c shows an example of such results. The curves shown as A, B, and C in Fig. 3a correspond to the term $A_+ Y(\Delta_{s+} r)$, the term $B_+ Y(\Delta_{p+} r)$, and $V_{\text{eff}}(r)$, respectively. These curves are obtained for $E_m = 1.0$ eV, $E_p = 1.2$ eV, and the detuning $\delta \equiv E_m - E_s < 0$, where both $A_\pm$ and $B_\pm$ are negative from (102a) and (102b), and thus the total potential $V_{\text{eff}}(r)$ is negative, that is, an attractive potential is formed. On the other hand, the curves denoted A, B, and C in Fig. 3b and c correspond to the terms $A_+ Y(\Delta_{s+} r) - A_- Y(\Delta_{s-} r)$, $B_+ Y(\Delta_{p+} r) - B_- Y(\Delta_{p-} r)$, and $V_{\text{eff}}(r)$, re-

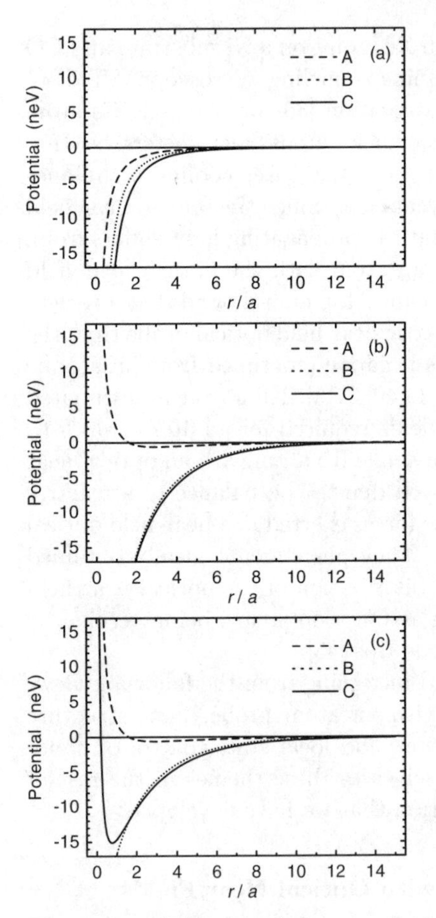

Fig. 3a–c. Examples of the near-field optical potential V_{eff} shown in the solid curves. The horizontal axis is normalized by the probe tip size a (see the text for details of the parameters used)

spectively. We use $E_m = 1.8$ eV, $E_p = 1.0$ eV, and $\delta > 0$ in Fig. 3b, while in Fig. 3c $E_m = 1.8$ eV, $E_p = 1.2$ eV, and $\delta > 0$. It follows from the figures that the potential value depends on the sign and magnitude of each term. In both figures, (b) and (c), the term $A_+ Y(\Delta_{s+} r) - A_- Y(\Delta_{s-} r)$ represented by curve A is negative, and thus an attractive potential is formed for the atom. The term $B_+ Y(\Delta_{p+} r) - B_- Y(\Delta_{p-} r)$ shown by curve B is positive, and results in a repulsive potential for the atom. Moreover, we have $|A_+ Y(\Delta_{s+} r) - A_- Y(\Delta_{s-} r)| > |B_+ Y(\Delta_{p+} r) - B_- Y(\Delta_{p-} r)|$ for Fig. 3(b), and as a result the total potential $V_{\text{eff}}(r)$ becomes attractive. In contrast to Fig. 3b, the potential changes sign in Fig. 3c when r goes beyond some value. Since $|A_+ Y(\Delta_{s+} r) - A_- Y(\Delta_{s-} r)| > |B_+ Y(\Delta_{p+} r) - B_- Y(\Delta_{p-} r)|$ at large r, while $|A_+ Y(\Delta_{s+} r) - A_- Y(\Delta_{s-} r)| < |B_+ Y(\Delta_{p+} r) - B_- Y(\Delta_{p-} r)|$ at small r, the total potential $V_{\text{eff}}(r)$ has a minimum at some r and becomes a potential well. This example indicates that one can control the r-dependence of $V_{\text{eff}}(r)$, i.e., the shape of the near-field optical potential, by choosing appropriate materials and probe structure. In addition, we can expect a trapping potential for a single Rb atom around a location with the dimensions of a probe tip.

One of the important factors in experiments with a near-field optical probe is how small a probe tip should be used. Therefore one should examine the change in the near-field optical potential described by $V_{\text{eff}}(r)$ in (104a), taking account of the size of the probe tip. In the following we assume for simplicity that an atom is the point-like object with discrete energy levels, while the probe tip is a sphere with radius a. The tip sphere at the position of $\boldsymbol{r}_p$, as shown in Fig. 4, produces a near-field optical potential at the atom position $\boldsymbol{r}_{\text{A}}$:

$$\begin{aligned} V(r) &= \frac{1}{4\pi a^3/3} \int V_{\text{eff}}(|\boldsymbol{r}_{\text{A}} - (\boldsymbol{r}' + \boldsymbol{r}_p)|)\, \mathrm{d}^3 r' \\ &= \frac{\mu_s \mu_p E_{\text{pol}}^2}{(\hbar c)^2 a^3} \sum_{G=A}^{B} \sum_{g=s}^{p} \sum_{j=\pm} \frac{j G_j}{\Delta_{g\,j}^3} \\ &\quad \times \{(1 + a\Delta_{g\,j}) \exp(-\Delta_{g\,j} a) - (1 - a\Delta_{g\,j}) \exp(\Delta_{g\,j} a)\} Y(\Delta_{g\,j} r) \\ &\equiv \sum_{G=A}^{B} \sum_{g=s}^{p} \sum_{j=\pm} j Z_{g\,j} G_j Y(\Delta_{g\,j} r)\ , \end{aligned} \tag{105}$$

where we assume the Yukawa sources are homogeneously distributed within the probe-tip sphere. Thus the total potential $V(r)$ is expressed in terms of the Yukawa potential as a function of the distance between the center of the tip sphere and the atom position.

Now let us consider whether or not the total potential $V(r)$ can have a potential minimum suitable for atom trapping. As a test case, we again use a ^{85}Rb atom with $E_s = 1.59$ eV and $\mu_s = 7.5$ debye. A probe tip with a radius of 10 nm is assumed, whose transition dipole moment $\mu_p = 1.5$ debye and excitation energy $E_p = 1.51$ eV are assumed. They are coupled to macroscopic

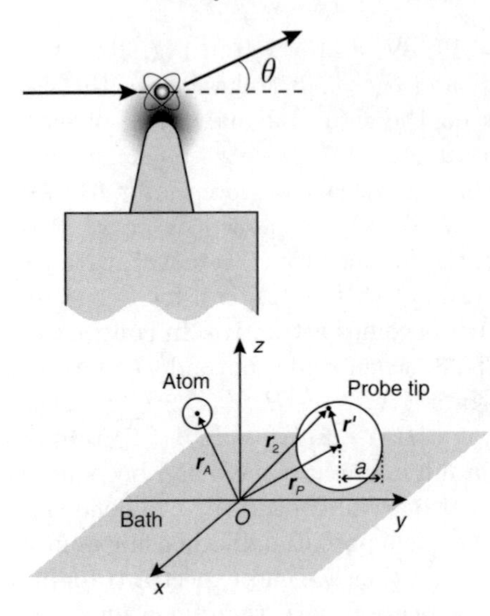

Fig. 4. Schematic drawing of an atom manipulation and the geometry of the model. The vectors $\boldsymbol{r}_{\mathrm{A}}$ and $\boldsymbol{r}_p$ denote the center of the atom and the center of the tip sphere with radius a, respectively. An arbitrary position inside the tip sphere is represented by $\boldsymbol{r}_2$ measured from the origin of the coordinate system, and $\boldsymbol{r}'$ measured from the center of the tip sphere, respectively

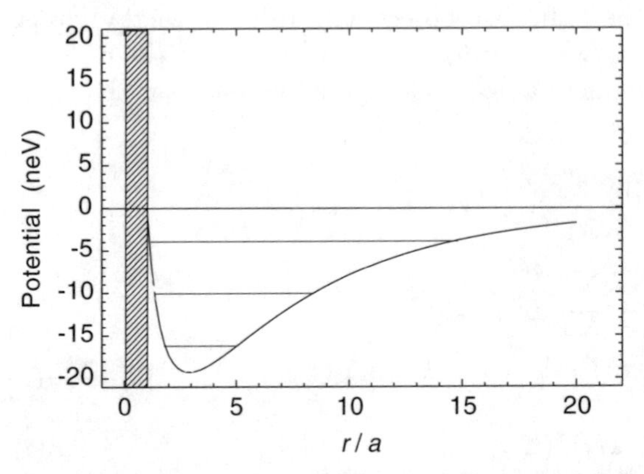

Fig. 5. Example of near-field optical potential for ^{85}Rb (resonance energy of 1.59 eV), represented by the solid curve. The probe tip is assumed to have a radius of 10 nm and an electronic excitation energy of 1.51 eV. Approximating it by a harmonic oscillator potential around the minimum point, three vibrational levels are shown. The shaded area shows the probe-tip size

matter with $E_m = 2.0$ eV. For the effective mass of the exciton polaritons, $E_{\mathrm{pol}} = m_{\mathrm{pol}}c^2 = E_m$ is employed. It follows from Fig. 5 that the potential has a minimum of -20 neV near a position $r = 2a$ from the probe-tip surface.

Approximating it by a harmonic oscillator potential around the minimum point, we find that two or three vibrational levels can be supported; the lowest vibrational energy with the label $n = 0$ corresponds to 3.1 neV, or equivalently 35 μK. This result suggests the possibility of single Rb-atom trapping at this level [19].

It may be interesting to compare the depth of the above potential well with the ones obtained for an atom and a microsphere system in which the radius of the sphere is much greater than ours. The potential depth for a Rb atom semi-classically calculated in [44] is of the same order of magnitude or possibly a little shallower than our results. However, their minimum position of the potential depends on the wavelength used and is different from ours. This is because they use a microsphere with a dielectric constant of 6 and a radius of about 1 μm for visible light. The potential for a Cs atom quantum-mechanically calculated with a 50-micron sphere shows the similar tendency [45]. It follows that the nonresonant term neglected in [44], that is, the optical near field leading to size- dependent effects becomes important as the tip sphere becomes smaller. The effect of an ideally conducting conical surface on an atom was also classically estimated [46], and the energy shift of the atom in the region of interest seems to be similar in order of magnitude, although the details of the parameters used are not known.

Atom Deflection by Optical Near Fields

The incident atomic beam for this kind of experiment is cooled by laser cooling techniques [47], so that the atomic velocity is approximately 1 m/s in vacuum, or equivalently 10 mK. Such atoms with low velocity are deflected or scattered by the near-field optical potential given by (105). This thus can be formulated as a potential scattering problem within the Born approximation. As the velocity of the atom decreases, the first Born approximation becomes invalid. It is known, however, that the first Born approximation breaks down when the incident velocity is close to the sub-meter-per-second range or the kinetic energy is in the submillikelvin range [48].

The differential scattering cross section $\sigma(\theta)$ is given in the first Born approximation by

$$\sigma(\theta) = \left| -\frac{1}{4\pi}\left(\frac{8\pi M}{K\hbar^2}\right)\int_a^\infty r dr V(r)\sin(Kr)\right|^2 , \tag{106}$$

where the atomic mass is denoted by M, and the momentum transfer K is defined in terms of the velocity of the atom v, scattering angle θ, and M as

$$K = 2\frac{Mv}{\hbar}\sin\left(\frac{\theta}{2}\right) . \tag{107}$$

Substituting the potential $V(r)$ described by (105) into (106), we have an explicit formula for the differential scattering cross section [19]:

$$\sigma(\theta) = \left| \left(\frac{2M}{K\hbar^2}\right) \sum_{G=A}^{B} \sum_{g=s}^{p} \sum_{j=\pm} jZ_{gj}G_j \int_a^\infty \mathrm{d}r \exp(-\Delta_{gj}r) \sin(Kr) \right|^2$$

$$= \left| \left(\frac{2M}{K\hbar^2}\right) \sum_{G=A}^{B} \sum_{g=s}^{p} \sum_{j=\pm} jZ_{gj}G_j \frac{K\cos(Ka) + \Delta_{gj}\sin(Ka)}{K^2 + \Delta_{gj}^2} \right|^2, \quad (108)$$

which provides us with the ratio of the atomic flux at scattering angle θ to the incident atomic flux.

Figure 6a and b shows the differential scattering cross section of ^{85}Rb with an incident velocity of 1 m/s when the effective potential $V_{\text{eff}}(r)$ is the attractive potential represented by curve C in Fig. 3a and b. The curves denoted A, B, and C are calculated with a tip sphere radius of 10, 30, and 50 nm, respectively. The periodic structure in both figures results from the finite size of the probe tip. From the analytic expression it follows that the periodic length is inversely proportional to the tip size: the larger the tip size, the shorter the period. Comparing with the first minimum of the differential scattering cross section, we see that a smaller probe tip can deflect the atom more strongly in both figures, although the optimum size should be determined from a discussion of the de Broglie wavelength of the atom. The difference between Figs. 6a and b can be understood from the sign of each term in $V(r)$ given by (105). In Fig. 6a, each component in (108) has the same sign in terms of the sum of $Z_{gj}G_j$ and j, and constructively contributes to the scattering amplitude. On the other hand, each component in (108) has the opposite sign for Fig. 6b, and is destructively summed in the scattering amplitude. This reduces the first minimum of the deflection angle θ.

In order to estimate a typical displacement of the atomic beam from the incident direction, i.e., a typical deflection angle, let us assume a case in which the incident atomic flux N is $10^{10}/(\text{cm}^2\,\text{s})$ and the atomic flux required for detection N_d is 10^3/s. It is necessary for the measurement data to be meaningful that the inequality $N\sigma(\theta) \geq N_d$ holds. It follows from curve A in Fig. 6b that this condition is satisfied if $\theta \leq 1°$ with a tip radius of 10 nm. This deflection angle is large enough to be detected in the current experimental situation.

7.2 Fundamental Properties of Optical Near-Field Microscopy

In this section, we show that our formulation is also applicable to conventional problems in optical near fields in a unified way, as well as atom manipulation discussed above. The optical near-field intensity in an idealized probe–sample system is calculated in order to analyze fundamental properties of detected signals of optical near-field microscopy [49].

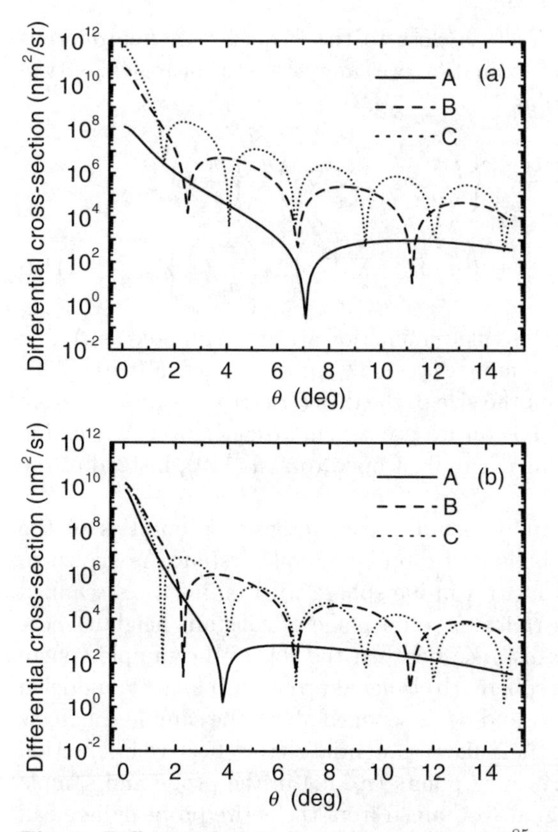

Fig. 6. Differential scattering cross section of ^{85}Rb with an incident velocity of 1 m/s in the first Born approximation. The radius of the tip is 10 nm (A), 30 nm (B), or 50 nm (C). In (**a**) the near-field optical potential shown in Fig. 3a is assumed, while the potential corresponding to Fig. 3b is assumed in (**b**)

As shown in the preceding section, virtual exciton polariton exchange is a source of the effective probe–sample interaction. From the dispersion relation of the exciton polaritons, it follows that they are massive, with an effective mass which is usually determined from $\hbar k \left|\mathrm{d}\Omega(k)/\mathrm{d}k\right|_{k=0}^{-1}$. Here it should be noted that the electrons in a nanometric probe or sample are locally confined and that electron wave numbers are also localized around $k_0 = \pi/a_\alpha$, depending on the size a_α ($\alpha = s, p$) of the probe–sample system. Then the wave number or momentum of the exciton polariton virtually exchanged between the probe and sample is expected to close to the wave number k_0. Thus we define the effective mass of the exciton polaritons at $k = k_0$ instead of $k = 0$:

$$\frac{1}{m_{\mathrm{pol}}} = \left[\frac{1}{\hbar k}\left|\frac{\mathrm{d}\Omega(k)}{\mathrm{d}k}\right|\right]_{k=k_0} \approx \frac{c}{\hbar k_0} = \frac{a_\alpha c}{\pi\hbar}\,, \tag{109}$$

because the eigenfrequency $\Omega(k)$ is close to the frequency of free photons around $k = k_0$. Using this approximation, we can rewrite the near-field optical potential described in (104a) as

$$V_{\text{eff}}(r) = \sum_{i,j=1}^{3} \frac{8(\mu_s)_i(\mu_p)_j}{9} \left\{ \left(\frac{\pi^2}{a_s^2}\delta_{ij} - \nabla_i\nabla_j \right) Y\left(\frac{\pi}{a_s}r\right) + \left(\frac{\pi^2}{a_p^2}\delta_{ij} - \nabla_i\nabla_j \right) Y\left(\frac{\pi}{a_p}r\right) \right\}, \tag{110}$$

where the effective masses of the Yukawa function are approximated as $\Delta_{s\pm} = \pi/a_s$ and $\Delta_{p\pm} = \pi/a_p$ by assuming $\Omega_0(s)$, $\Omega_0(p) \ll ck_0$ (cf. (101b)). This assumption is reasonable when the size of the dielectric probe–sample system is a few tens of nanometers. For future use we sum exactly over λ, yielding the additional terms consisting of gradient operators in (110), instead of the previous averaging to 2/3.

Now let us discuss the fundamental signal intensity properties of the microscopy, using a typical nanometric probe–sample system. As shown in Fig. 7, we consider two cases: (a) a probe sphere with radius a_1 is scanned above a sample sphere with radius a_s on a plane of constant height, where the height h is the shortest distance between the top of the sample sphere and the bottom of the probe sphere; (b) tapered probe with angle θ, modeled by two spheres with radii of a_1 and a_2, is scanned above the sample sphere as in the first case. From (110), the Yukawa potential as the effective interaction is generated between two arbitrary points $(\boldsymbol{r}_i, \boldsymbol{r}_s)$ in the probe and sample spheres. Thus, the pickup signal I obtained from the entire probe sphere can

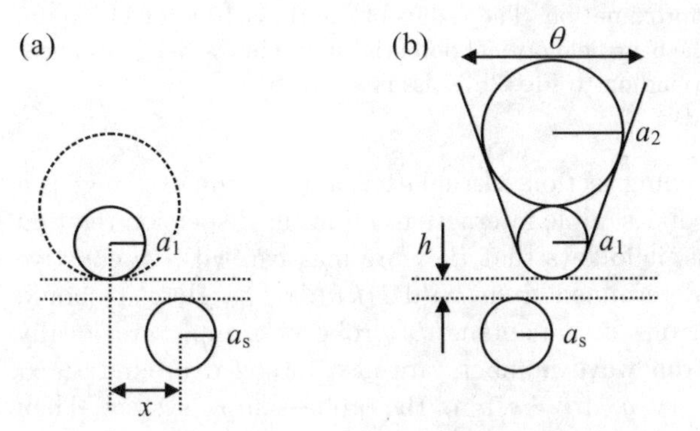

Fig. 7. Illustration of the models for an optical near-field probe and sample system: **(a)** a spherical probe with radius a_1, and **(b)** a tapered probe consisting of two spheres with radii of a_1 and a_2

be written

$$
\begin{aligned}
I(r_{sp}) &\equiv \sum_{i=1}^{2} \left| \int \int \nabla_{r_i} V_{\text{eff}} \left(|\boldsymbol{r}_i - \boldsymbol{r}_s| \right) \mathrm{d}^3 r_s \mathrm{d}^3 r_i \right|^2 \\
&\propto \sum_{i=1}^{2} \left[a_i^3 \left\{ \frac{a_s}{a_i} \cosh\left(\frac{\pi a_s}{a_i} \right) - \frac{1}{\pi} \sinh\left(\frac{\pi a_s}{a_i} \right) \right\} \right. \\
&\quad \left. \times \left(\frac{1}{r_{sp}} + \frac{a_i}{\pi r_{sp}^2} \right) \exp\left(-\frac{\pi r_{sp}}{a_i} \right) \right]^2 , \qquad (111)
\end{aligned}
$$

where r_{sp} denotes the distance between the center of the probe and the sample. The pickup signal decays in a similar way to the Yukawa function. The numerical results based on (111) are presented for the case (a) in Figs. 8 and 9, and for the case (b) in Figs. 10 and 11.

Figure 8 shows the signal intensity from the sample with $a_s = 10$ nm detected by the probe with $a_1 = 10$ nm or 20 nm, scanned in the lateral direction with constant height $h = 1$ nm. It follows that each full width at half maximum determining the lateral resolution of the system is nearly equal to the size of each probe tip. In Fig. 9, we present the signal intensity normalized by the volume of both probe and sample spheres when the probe radius is varied with the radius of the sample sphere fixed at $a_s = 10$ nm or 20 nm. It shows that the signal remains the highest when the probe size is comparable to the sample size. This is called the size-resonance effect. These results are consistent with numerical ones obtained with other methods [50,51].

In Fig. 10, we plot the signal intensity as a function of the probe position, varying the taper angle θ. When the taper angle is as small as $\theta = 20°$, as shown in Fig. 10a, the signal intensity due to the tapered part is negligibly small and the net signal is determined by the apex contribution. This means that the lateral resolution is dominated by the apex part. By contrast, as the

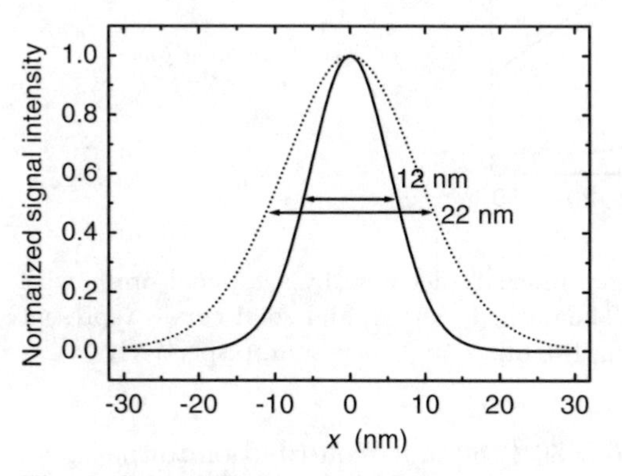

Fig. 8. Spatial distribution of signal intensity scanned with probe-tip sizes of $a_1 =$ 10 nm (*solid curve*) and 20 nm (*dotted curve*), where the sample size $a_s = 10$ nm is assumed

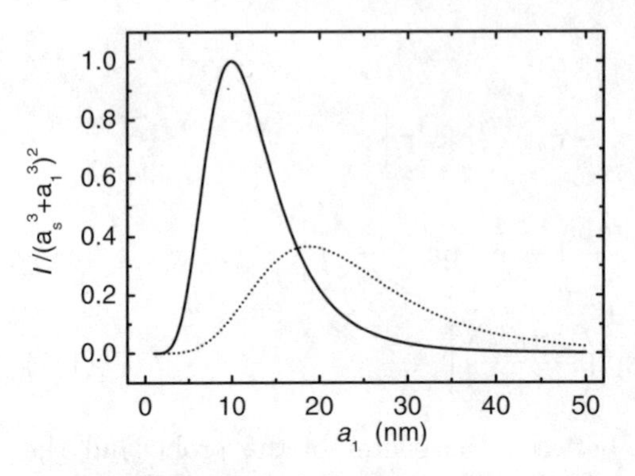

Fig. 9. Dependence of signal intensity on probe-tip size. The solid and dotted curves represent results for sample sizes $a_s = 10$ nm and 20 nm, respectively

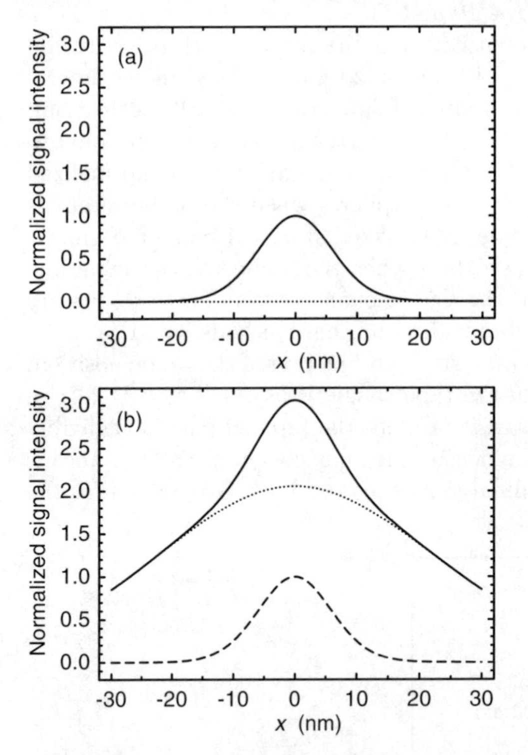

Fig. 10. Spatial distribution of signal intensity detected by a tapered probe with taper angles (**a**) 20° and (**b**) 80°. The dashed, dotted, and solid curves represent the contributions from the probe tip, the taper, and their sum, respectively

taper angle becomes as large as $\theta = 80°$, the taper contribution to the signal intensity becomes significant. This broadens the width of the net signal (see Fig. 10b), leading to the degradation of contrast, that is, the ratio of the

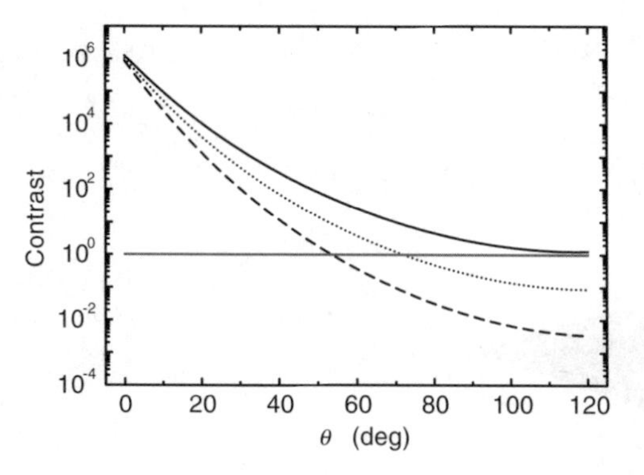

Fig. 11. Variation of the ratio of signal intensity from the apex part to that from the tapered part as a function of taper angle. The solid, dotted, and dashed curves represent results for heights, $h = 1$ nm, 5 nm, and 10 nm, respectively. The region above the horizontal line, the ratio 1, indicates that the contrast can be determined from the probe tip, not the taper part. By contrast, the tapered part governs the contrast in the region below the ratio 1

signal intensity from the apex part to that from the tapered part. Figure 11 shows the dependence of this contrast on taper angle θ. It illustrates how the approach distance of the probe affects the contrast of the image. For an approach distance as small as $h = 1$ nm, the interaction between the apex and sample spheres is stronger than that between the taper and sample spheres. Thus, the apex sphere is the main factor determining the contrast, as well as the lateral resolution. As the approach distance becomes as large as $h = 10$ nm, the interaction between the apex and the sample is weakened in contrast with that between the taper and the sample. This can be understood by noting that the interaction range of a larger sphere is longer than that of a smaller sphere. Therefore, in such a case, the tapered part gradually determines the contrast with increasing taper angle. At taper angles as large as 60°, the contrast degrades again. It is worth noting here that we could improve the contrast, and accordingly the lateral resolution of the system, by screening the tapered part with some kind of metallic coating, which would reduce the interaction between the taper part and the sample.

At the end of this section, we comment on the polarization dependence of the signal intensity. It follows from (110) that the polarization effect manifests itself near the edge of a sample. As an example, we show the signal intensity from a circular aperture in Fig. 12, where the incident polarization relative to the aperture is fixed in the x-direction, while all polarization components of the probe are detected. The signal intensity is enhanced near both sides of the aperture that are perpendicular to the incident polarization direction. In Fig. 12b the dependence of the signal intensity on approach distance of the probe is plotted. The enhancement of the signal at the aperture edges disappears, as the probe is located at a higher position. Such polarization

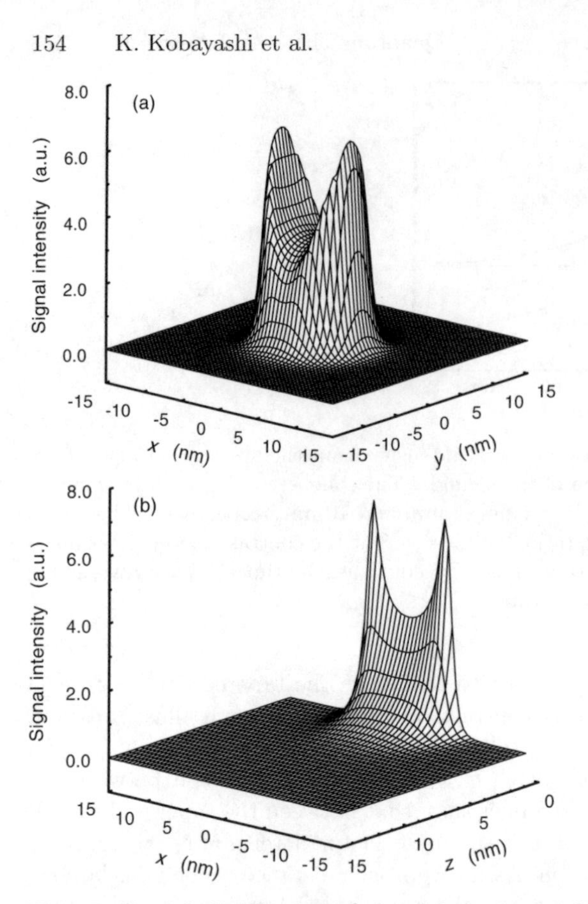

Fig. 12. Spatial distribution of signal intensity in the (**a**) xy-plane and (**b**) xz-plane. The circular aperture with a radius of 10 nm is chosen as a sample, and the polarization of incident light is parallel to the x-axis

dependence of the signal intensity has been explained theoretically and numerically by using macroscopic electromagnetic theory. Exact expressions of electromagnetic fields have been obtained for an infinitely thin conducting plane with a small aperture by Leviatan [52], and for a relatively thick one by Roberts [53]. Novotny et al. [54] have shown similar results numerically by means of the multiple multipole (MMP) method. The results presented here are qualitatively consistent with these previous studies.

8 Outlook

From both basic and application viewpoints, there remain in this field a lot of intriguing issues that we could not discuss. The emphasis here lies in the coupling between a nanometric subsystem n and a macroscopic subsystem M, and the effective interaction induced in the nanometric subsystem. In such a system, cavity quantum electrodynamics (QED) effects are to be expected. One of them is the Casimir–Polder effect [55] due to vacuum fluctuation and retardation effects, both of which are easily included in our framework. In addition, cooperative or destructive effects inherent in the system might be found in dissociation and adsorption processes of atoms or molecules, which will help us to elucidate the nature of nanostructure and take advantage of it.

It will be possible to have another formulation emphasizing a "dressed photon" picture if the P space is spanned by including degrees of freedom of massive virtual photons mediating between constituents of a nanometric subsystem. Such a formulation will reveal more about the difference between propagating real photons and massive virtual photons, phases, and numbers of tunneling photons. Moreover, it can provide new understanding and insights related to observation problems and the reaction mechanism.

A low-dimensional system such as a nanometric quantum dot system is another candidate for a sample interacting with an optical near field as, well as an atom or a molecule. The competition of several coupling sources may be of principle importance in such a system. There are fundamental and profound problems when we regard the optical near field as a mixed state of photons and matter excitations: what is the basic equation governing coherence and decoherence of the system? How are the coherence, population, and excitation energy transferred from one dot to another? The answers to such problems will open up new possibilities of applications using the optical near field such as nanometric photonics devices based on a new principle of operation.

References

1. J.D. Jackson: *Classical Electrodynamics*, 3rd edn. (Wiley, New York 1999)
2. J.P. Fillard: *Near Field Optics and Nanoscopy* (World Scientific, Singapore 1996)
3. M.A. Paesler, P.J. Moyer: *Near-Field Optics* (John Wiley & Sons, New York 1996)
4. K. Cho: Prog. Theor. Phys. Suppl. **106**, 225 (1991)
5. H. Ishihara, K. Cho: Phys. Rev. B **48**, 7960 (1993)
6. K. Cho, Y. Ohfuti, K. Arima: Surf. Sci. **363**, 378 (1996)
7. M. Ohtsu, H. Hori: *Near-Field Nano-Optics* (Kluwer Academic /Plenum Publishers, New York 1999) pp. 281–296
8. H. Hori: In: *Near Field Optics* ed. by D.W. Pohl, D. Courjon (Kluwer Academic, Dordrecht 1993) pp. 105–114
9. K. Kobayashi, M. Ohtsu: In: *Near-Field Nano/Atom Optics and Technology* ed. by M. Ohtsu (Springer, Tokyo Berlin Heidelberg 1998) pp. 288–290
10. J.J. Sakurai: *Advanced Quantum Mechanics* (Addison-Wesley, Reading 1967)
11. U. Weiss: *Quantum Dissipative Systems*, 2nd edn. (World Scientific, Singapore 1999)
12. P. Fulde: *Electron Correlations in Molecules and Solids*, 2nd edn. (Springer, Berlin Heidelberg New York 1993)

13. H. Grabert: *Projection Operator Techniques in Nonequilibrium Statistical Mechanics* (Springer, Berlin Heidelberg New York 1982)
14. F. Haake: *Statistical Treatment of Open Systems by Generalized Master Equations* (Springer, Berlin Heidelberg New York 1973)
15. C.R. Willis, R.H. Picard: Phys. Rev. A **9**, 1343 (1974)
16. J. Rau, B. Muller: Phys. Rep. **272**, 1 (1996)
17. H. Hyuga, H. Ohtsubo: Nucl. Phys. **A294**, 348 (1978)
18. K. Kobayashi, M. Ohtsu: J. Microscopy **194**, 249 (1999)
19. K. Kobayashi, S. Sangu, H. Ito, M. Ohtsu: Phys. Rev. A **63**, 013806 (2001)
20. A.L. Fetter, J.D. Walecka: *Quantum Theory of Many-Particle Systems* (McGraw-Hill, New York 1971)
21. C. Cohen-Tannoudji, J. Dupont-Roc, G. Grynberg: *Photons and Atoms* (John Wiley & Sons, New York 1989)
22. C. Cohen-Tannoudji, J. Dupont-Roc, G. Grynberg: *Atom–Photon Interactions* (John Wiley & Sons, New York 1992)
23. D.P. Craig, T. Thirunamachandran: *Molecular Quantum Electrodynamics* (Dover, New York 1998)
24. M. Scully, M.S. Zubairy: *Quantum Optics* (Cambridge University Press, Cambridge 1997)
25. M. Kaku: *Quantum Field Theory* (Oxford University Press, Oxford 1993) pp. 295–320
26. S. Weinberg: *The Quantum Theory of Fields I, II, and III* (Cambridge University Press, Cambridge 1995)
27. H. Haken: *Quantum Field Theory of Solids* (North-Holland, Amsterdam 1983)
28. J.J. Hopfield: Phys. Rev. **112**, 1555 (1958)
29. K. Cho: J. Phys. Soc. Jpn. **55**, 4113 (1986)
30. C. Kittel: *Quantum Theory of Solids* (John Wiley & Sons, New York 1972)
31. D. Pines: *Elementary Excitations in Solids* (Perseus Books, Reading 1999)
32. P.W. Anderson: *Concepts in Solids* (World Scientific, Singapore 1997)
33. C. Cohen-Tannoudji: *Atoms in Electromagnetic Fields* (World Scientific, Singapore 1994)
34. M. Ohtsu, K. Kobayashi, H. Ito, G.-H. Lee: Proc. IEEE **88**, 1499 (2000)
35. D.M. Eigler, E.K. Schweizer: Nature **344**, 524 (1990)
36. P. Zeppenfeld et al.: Ultramicroscopy **42–44**, 128 (1992)
37. U. Staufer: In: *Scanning Tunneling Microscopy II* ed. by R. Wiesendanger, H.-J. Güntherodt (Springer, Berlin Heidelberg New York 1992) pp. 273–302
38. J. Ye, D.W. Vernooy, H.J. Kimble: Phys. Rev. Lett. **83**, 4987 (1999)
39. M. Ohtsu, S. Jiang, T. Pangaribuan, M. Kozuma: In: *Near Field Optics* ed. by D.W. Pohl, D. Courjon (Kluwer Academic, Dordrecht 1993) pp. 131–139
40. J.P. Dowling, J. Gea-Banacloche: In: *Advances in Atomic, Molecular, and Optical Physics 37* ed. by B. Bederson, H. Walther (Academic Press, San Diego 1996) pp. 1–94
41. V.V. Klimov, V.S. Letokhov: Opt. Commun. **121**, 130 (1995)
42. H. Ito, K. Otake, M. Ohtsu: Proc. SPIE **3467**, 250 (1998)
43. H. Ito, A. Takamizawa, H. Tanioka, M. Ohtsu: Proc. SPIE **3791**, 2 (1999)
44. V. Klimov, V.S. Letokhov, M. Ducloy: Eur. Phys. J. D **5**, 345 (1999)
45. D.W. Vernooy, H.J. Kimble: Phys. Rev. A **55**, 1239 (1997)
46. V.V. Klimov, Y.A. Perventsev: Quantum Electronics **29**, 847 (1999)

47. For example, *Laser Cooling and Trapping of Atoms* ed. by S. Chu, C. Wieman: J. Opt. Soc. Am. B **6**, 2020 (1989)
48. K. Kobayashi, S. Sangu, H. Ito, M. Ohtsu: In: *Near-Field Optics: Principles and Applications* ed. by X. Zhu, M. Ohtsu (World Scientific, Singapore 2000) pp. 82–88
49. S. Sangu, K. Kobayashi, M. Ohtsu: J. Microscopy **202**, 279 (2001)
50. K. Jang, W. Jhe: Opt. Lett. **21**, 236 (1996)
51. T. Saiki, M. Ohtsu: In: *Near-Field Nano/Atom Optics and Technology* ed. by M. Ohtsu (Springer, Tokyo Berlin Heidelberg 1998) pp. 15–29
52. Y. Leviatan: J. Appl. Phys. **60**, 1577 (1986)
53. A. Roberts: J. Appl. Phys. **70**, 4045 (1991)
54. L. Novotny, D.W. Pohl, P. Regli: J. Opt. Soc. Am. A **11**, 1768 (1994)
55. For example, *Cavity Quantum Electrodynamics* ed. by P.R. Berman (Academic Press, San Diego 1994)

Index

Springer Series in

OPTICAL SCIENCES

New editions of volumes prior to volume 60

1 **Solid-State Laser Engineering**
By W. Koechner, 5th revised and updated ed. 1999, 472 figs., 55 tabs., XII, 746 pages

14 **Laser Crystals**
Their Physics and Properties
By A. A. Kaminskii, 2nd ed. 1990, 89 figs., 56 tabs., XVI, 456 pages

15 **X-Ray Spectroscopy**
An Introduction
By B. K. Agarwal, 2nd ed. 1991, 239 figs., XV, 419 pages

36 **Transmission Electron Microscopy**
Physics of Image Formation and Microanalysis
By L. Reimer, 4th ed. 1997, 273 figs. XVI, 584 pages

45 **Scanning Electron Microscopy**
Physics of Image Formation and Microanalysis
By L. Reimer, 2nd completely revised and updated ed. 1998,
260 figs., XIV, 527 pages

Published titles since volume 60

60 **Holographic Interferometry in Experimental Mechanics**
By Yu. I. Ostrovsky, V. P. Shchepinov, V. V. Yakovlev, 1991, 167 figs., IX, 248 pages

61 **Millimetre and Submillimetre Wavelength Lasers**
A Handbook of cw Measurements
By N. G. Douglas, 1989, 15 figs., IX, 278 pages

62 **Photoacoustic and Photothermal Phenomena II**
Proceedings of the 6th International Topical Meeting, Baltimore, Maryland,
July 31 - August 3, 1989
By J. C. Murphy, J. W. Maclachlan Spicer, L. C. Aamodt, B. S. H. Royce (Eds.),
1990, 389 figs., 23 tabs., XXI, 545 pages

63 **Electron Energy Loss Spectrometers**
The Technology of High Performance
By H. Ibach, 1991, 103 figs., VIII, 178 pages

64 **Handbook of Nonlinear Optical Crystals**
By V. G. Dmitriev, G. G. Gurzadyan, D. N. Nikogosyan,
3rd revised ed. 1999, 39 figs., XVIII, 413 pages

65 **High-Power Dye Lasers**
By F. J. Duarte (Ed.), 1991, 93 figs., XIII, 252 pages

66 **Silver-Halide Recording Materials**
for Holography and Their Processing
By H. I. Bjelkhagen, 2nd ed. 1995, 64 figs., XX, 440 pages

67 **X-Ray Microscopy III**
Proceedings of the Third International Conference, London, September 3-7, 1990
By A. G. Michette, G. R. Morrison, C. J. Buckley (Eds.), 1992, 359 figs., XVI, 491 pages

68 **Holographic Interferometry**
Principles and Methods
By P. K. Rastogi (Ed.), 1994, 178 figs., 3 in color, XIII, 328 pages

69 **Photoacoustic and Photothermal Phenomena III**
Proceedings of the 7th International Topical Meeting, Doorwerth, The Netherlands,
August 26-30, 1991
By D. Bicanic (Ed.), 1992, 501 figs., XXVIII, 731 pages

Springer Series in

OPTICAL SCIENCES

70 **Electron Holography**
By A. Tonomura, 2nd, enlarged ed. 1999, 127 figs., XII, 162 pages

71 **Energy-Filtering Transmission Electron Microscopy**
By L. Reimer (Ed.), 1995, 199 figs., XIV, 424 pages

72 **Nonlinear Optical Effects and Materials**
By P. Günter (Ed.), 2000, 174 figs., 43 tabs., XIV, 540 pages

73 **Evanescent Waves**
From Newtonian Optics to Atomic Optics
By F. de Fornel, 2001, 277 figs., XVIII, 268 pages

74 **International Trends in Optics and Photonics**
ICO IV
By T. Asakura (Ed.), 1999, 190 figs., 14 tabs., XX, 426 pages

75 **Advanced Optical Imaging Theory**
By M. Gu, 2000, 93 figs., XII, 214 pages

76 **Holographic Data Storage**
By H.J. Coufal, D. Psaltis, G.T. Sincerbox (Eds.), 2000
228 figs., 64 in color, 12 tabs., XXVI, 486 pages

77 **Solid-State Lasers for Materials Processing**
Fundamental Relations and Technical Realizations
By R. Iffländer, 2001, 230 figs., 73 tabs., XVIII, 350 pages

78 **Holography**
The First 50 Years
By J.-M. Fournier (Ed.), 2001, 266 figs., XII, 460 pages

79 **Mathematical Methods of Quantum Optics**
By R.R. Puri, 2001, 13 figs., XIV, 285 pages

80 **Optical Properties of Photonic Crystals**
By K. Sakoda, 2001, 95 figs., 28 tabs., XII, 223 pages

81 **Photonic Analog-to-Digital Conversion**
By B.L. Shoop, 2001, 259 figs., 11 tabs., XIV, 330 pages

82 **Spatial Solitons**
By S. Trillo, W.E. Torruellas (Eds), 2001, 194 figs., 7 tabs., XX, 454 pages

83 **Nonimaging Fresnel Lenses**
Design and Performance of Solar Concentrators
By R. Leutz, A. Suzuki, 2001, 139 figs., 44 tabs., XII, 272 pages

84 **Nano-Optics**
By S. Kawata, M. Ohtsu, M. Irie (Eds.), 2002, 258 figs., 2 tabs., XVI, 321 pages

85 **Sensing with Terahertz Radiation**
By D. Mittleman (Ed.), 2003, 207 figs., 14 tabs., XIV, 334 pages

86 **Progress in Nano-Electro-Optics I**
Basics and Theory of Near-Field Optics
By M. Ohtsu (Ed.), 2003, 118 figs., XIV, 163 pages

Printing (Computer to Film): Saladruck Berlin
Binding: Stürtz AG, Würzburg